ODD-ELECTRON SPECIES

ODD-ELECTRON SPECIES

The Chemistry of Free Radicals

P. R. SCOTT

Charterhouse, Godalming

Cambridge University Press

Cambridge
London New York New Rochelle
Melbourne Sydney

Published by the Press Syndicate of the University of Cambridge
The Pitt Building, Trumpington Street, Cambridge CB2 1RP
32 East 57th Street, New York, NY 10022, USA
296 Beaconsfield Parade, Middle Park, Melbourne 3206, Australia

First published 1981

Printed in Great Britain at the University Press, Cambridge

Scott, P. R.
Odd-electron species.
1. Radicals (Chemistry)
I. Title
541.2′24 QD471

ISBN 0 521 28177 6

Contents

Preface

The idea that chemical compounds contain ions or molecules which have complete electron shells is the cornerstone of our understanding of chemical valency; it allows us to predict the formulae and properties of known compounds, and to explain why other substances do not exist. It is an elegant concept of considerable power, but unfortunately it is also far from being perfect.

This book is concerned with ions and molecules which contain odd numbers of electrons, and which therefore have incomplete electron shells. These may be thought of as strange and unusual species, but in fact they are remarkably common, and of great importance in many areas of chemistry. I hope that this book may raise them from the level of being seen as difficult exceptions, which must simply be learnt parrot-fashion, for it is my contention that the subject is not only comprehensible, but contains some fascinating chemistry. Chapter 1 should be read first, but the remaining chapters are, as far as is possible, independent of each other, and so they may be read in any order.

The question of chemical nomenclature is still more vexed than some would wish to claim; I have used IUPAC names for the great majority of compounds in the book, but make no apology for occasional trivial names, such as nitric oxide, where such usage is widespread.

It would be impossible to thank all those who have helped, either wittingly or unwittingly, in the production of this book, but I must express my gratitude to the Master and Fellows of Selwyn College, Cambridge, for their splendid hospitality, and to Charterhouse School for their kindness in allowing me a sabbatical term.

Charterhouse,
Godalming.

P.R.S.

1 Odd-electron species

A Introduction

One of the most important questions which we can ask ourselves in chemistry is how and why elements combine to form compounds. Early scientists tackled this problem by conducting experiments to discover what mass of one element was required to react with a given mass of another element, and in this way they were able to find the empirical formulae of simple compounds. From these formulae the idea of valency, or combining power, arose, and they were able to assign a valency to each element. Thus zinc has a valency of two and chlorine one, so zinc chloride has the formula $ZnCl_2$.

When the valencies of a number of elements became known, it was clear that valency is related to position in the periodic table: the alkali metals all have valency one, the alkaline earth metals two, the halogens one, and so on. When the electronic structures of atoms became known, the variation of valency across the periodic table could be explained, and the importance of complete electron shells was recognised. We may summarise this simple valence theory as follows: when elements combine with each other, they do so in such a way as to build up complete electron shells. They may do this by transferring one or more electrons from one atom to another, in which case an ionic compound is formed, or alternatively they may share one or more pairs of electrons, in which case a covalent compound is formed. These processes are conventionally represented as in figure 1.1, which shows the formation of NaCl and H_2O from atoms of their elements.

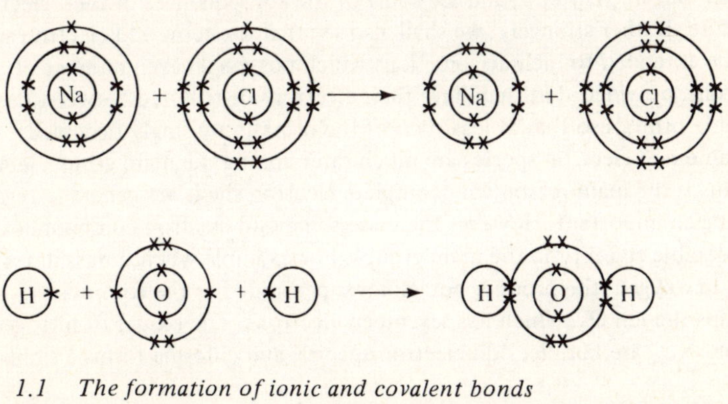

1.1 The formation of ionic and covalent bonds

As a theory this has much to commend it; it predicts with considerable accuracy the formulae of compounds formed by the combination of two elements, and also of those which we would not expect to be formed. It predicts correctly that a compound formed by a metal and non-metal should be ionic, whereas one formed from two non-metals should be covalent. It is extended simply to allow the prediction of shapes of small molecules: water is bent rather than linear, and ammonia pyramidal rather than planar. It is therefore scarcely surprising that the idea of the importance of complete electron shells has dominated our thinking about chemical compounds.

Unfortunately, although the successes of the complete electron shell theory are many, its failures cannot be ignored. It is not true to say that all chemical compounds have complete electron shells, and many important species have an odd number of electrons, which is 'not allowed' by simple valence theory. It is with these species that this book will be concerned; we shall need to look at the question of why substances with odd numbers of electrons are formed, and also what effect having an odd number of electrons has on their properties. There is no universally agreed term to describe a substance which has an odd number of electrons; the term *free radical* is commonly used to describe a neutral molecule with an odd number of electrons, such as CH_3, but it would not be normal to describe an ion with an odd number of electrons, such as O_2^-, in this way. We shall therefore use the term *odd-electron species* to describe any atom, molecule or ion which has an odd number of electrons.

B Stable odd-electron species

The most readily detectable odd-electron species are those which are stable indefinitely at room temperature. By far the largest group of compounds which fall into this category are those of the transition metals. Here the formation of ions with complete electron shells is the exception rather than the rule, and many common and important ions, such as Fe^{3+} and Cu^{2+}, have an odd number of electrons. We shall investigate the reasons for the stability of such ions in chapter 5, and see some of the consequences of their electronic structure. Rather strangely, we shall also see that the term odd-electron species may be extended to include some ions which possess an even number of electrons, provided that they have their electrons distributed between the available orbitals so that at least two of the orbitals are singly occupied.

Stable odd-electron species are much rarer among the main group elements, and this is the main reason why complete electron shells are generally regarded as being so important. However there are some odd-electron compounds of considerable stability in the main groups. For example when potassium is burnt in oxygen, the product potassium superoxide has the formula KO_2, and contains the ion O_2^-, which has seventeen electrons. The oxides of nitrogen NO and NO_2 are familiar odd-electron species, and chlorine forms a similar oxide ClO_2. These and other examples we shall discuss in chapters 4 and 7.

Odd-electron species which are stable indefinitely at room temperature clearly act as important tests of our understanding of how and why compounds are formed. But that is not the only reason why odd-electron species are so important in chemistry: many odd-electron species are also formed readily from stable molecules which have complete electron shells, by the action of heat, light or electron bombardment. As we shall see in this book, they act as important intermediates in chemical reactions and are often invaluable in analytical chemistry. They are certainly far too common to be written off as unimportant exceptions to more general rules, simply to be learnt by rote. In this book we shall learn to recognise odd-electron species, see the conditions under which they are produced, and then consider their physical and chemical properties.

C The experiments of Gomberg and Paneth

By the end of the nineteenth century it had been clearly established that the valency of carbon was four, and that it was invariant in all compounds with the single exception of carbon monoxide. Many attempts had been made to form compounds in which carbon showed other valencies, but it was generally agreed that none could exist.

A considerable stir was therefore caused by the results of some experiments by Gomberg, who in 1900 tried to produce the compound hexaphenylethane, $(C_6H_5)_3C{-}C(C_6H_5)_3$. The reaction of bromoalkanes with metals was known to produce symmetrical alkanes, and so he treated triphenylbromomethane, $(C_6H_5)_3C{-}Br$, with silver metal:

$$2Ph_3CBr + 2Ag = Ph_3C{-}CPh_3 + 2AgBr \quad (\text{where Ph} = C_6H_5)$$

He found that, if he carefully excluded oxygen from the reaction mixture, on removal of the solvent a solid was obtained whose composition corresponded to that expected for hexaphenylethane. Many substituted alkanes had already been prepared and characterised, and so the properties of hexaphenylethane could be predicted with some confidence: for example it would be expected to be colourless and chemically very unreactive. But Gomberg found that his product dissolved in benzene forming a yellow solution, and reacted readily with oxygen and iodine at room temperature:

$$Ph_3C{-}CPh_3 + O_2 = Ph_3C{-}O{-}O{-}CPh_3$$

and

$$Ph_3C{-}CPh_3 + I_2 = 2Ph_3C{-}I$$

After a series of investigations, he concluded that the properties of hexaphenylethane could be rationalised only if it was assumed that in solution it dissociated into two free radicals:

$$Ph_3C{-}CPh_3 = Ph_3C\cdot + \cdot CPh_3$$

Here the central covalent bond, which consists of a shared pair of electrons, splits leaving one electron on each carbon atom; the two radicals so formed are both electrically neutral. The odd electron on each carbon atom is represented by a single dot, in much the same way that a lone pair of electrons is sometimes represented by two dots. Notice that the carbon atom now has a valency of only three.

This was a key experiment, as it showed for the first time that a free radical containing a carbon atom could exist. Further work showed that Gomberg's conclusion was not exactly right, and that his yellow solution actually contained a mixture of free radicals and hexaphenylethane molecules in equilibrium with each other, with only about 5% radicals present. But that does not alter the significance of his discovery, and he later showed that if the benzene rings are replaced by larger aromatic systems, such as naphthalene rings, then the yield of radicals is 100%.

Gomberg's accidental discovery came some years before there was any understanding of the nature of covalent bonds, but we may now see the stability of the triphenylmethyl radical as arising from electron delocalisation. The single electron may be delocalised round all three benzene rings, giving rise to canonical structures such as:

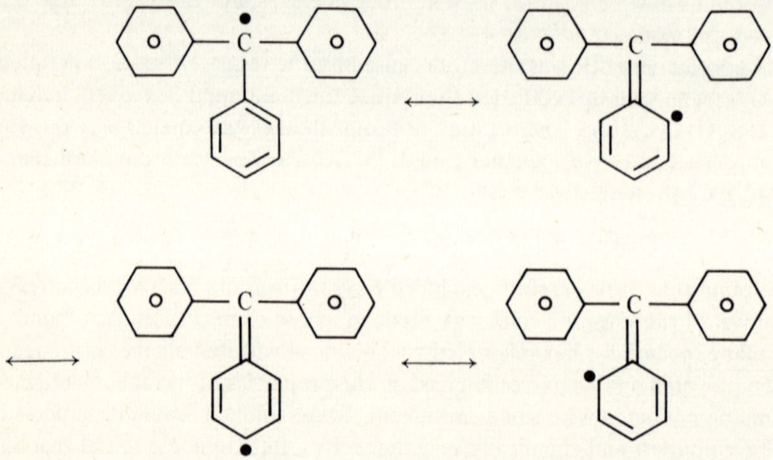

Gomberg's work had surprisingly little impact on the chemistry of his day. At first many chemists doubted the evidence for the triphenylmethyl radical, but even when its existence was established beyond doubt, it was generally regarded as a strange oddity and a single exception which had little significance or relevance to mainstream organic chemistry. It was not until 1929, when Paneth demonstrated the existence of the methyl radical $CH_3\cdot$, that it was understood that free radicals could be formed under mild conditions, and play an important role in chemical reactions.

Paneth's experiments were concerned with the thermal decomposition of lead tetramethyl, $Pb(CH_3)_4$. This decomposes on heating, depositing metallic lead and producing a mixture of gases, the main component of which is ethane. The reaction is irreversible as the product gases do not react with lead under mild conditions. Paneth was able to show that the decomposition involved methyl radicals using the apparatus of figure 1.2. A carrier gas such as helium or nitrogen is passed over liquid $Pb(CH_3)_4$, which evaporates and is

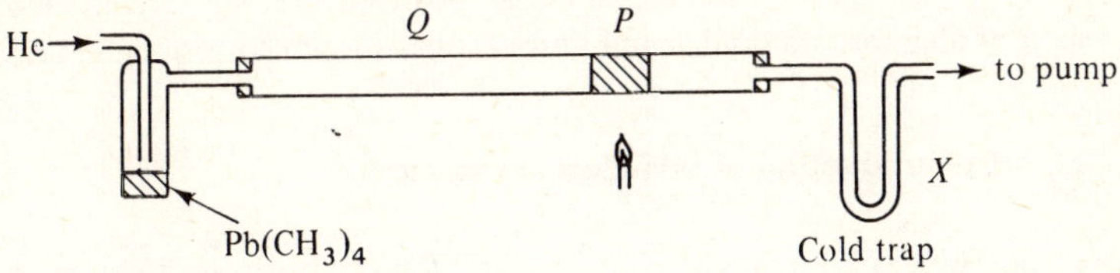

1.2 The Paneth experiment

transported down a glass tube. The tube is heated at point *P*, causing the lead tetramethyl to decompose, and a lead mirror is deposited on the inside of the glass. The heating is then transferred to point *Q*, further upstream than *P*. A new mirror is formed at *Q*, but Paneth was able to show that, provided *P* and *Q* are close together, the lead mirror at *P* is removed while the new mirror at *Q* is being deposited. If *P* and *Q* are widely separated, however, the lead at *P* is not removed. By placing a cold trap at point *X* Paneth was able to show that, when *P* and *Q* are close together, the lead mirror at *P* is converted back into lead tetramethyl.

Paneth interpreted his results in the following way: when the $Pb(CH_3)_4$ is heated, it decomposes giving lead metal and $CH_3\cdot$ radicals:

$$Pb(CH_3)_4 = Pb + 4CH_3\cdot$$

These methyl radicals contain just seven electrons in the outer shell of the carbon atom. Under normal circumstances they react with each other, forming mostly ethane:

$$CH_3\cdot + CH_3\cdot = CH_3{-}CH_3$$

This reaction is typical of alkyl radicals, whose chemistry we shall study in more detail in the next chapter. But in the Paneth experiment the methyl radicals can undergo a different reaction before they have time to react with each other:

$$Pb + 4CH_3\cdot = Pb(CH_3)_4$$

This is simply the reverse of the way in which they were formed: decomposition occurs on heating at *Q*, and recombination on cooling at *P*. Paneth's interpretation of the experiment seems the only sensible one: the decomposition of $Pb(CH_3)_4$ produces a reactive intermediate, and as this intermediate reacts with Pb forming $Pb(CH_3)_4$ only, and with itself forming $CH_3{-}CH_3$, it is clearly $CH_3\cdot$.

By measuring the flow rate, and the maximum distance between *P* and *Q* which allows the lead mirror to be removed, it is possible to find the average lifetime of the $CH_3\cdot$ radicals before they combine with each other to form ethane. Paneth showed that with a gas flow rate of 15 m s^{-1}, the maximum separation of *P* and *Q* is about 30 cm, giving a lifetime for the radicals of about 0.02 s.

The use of metallic mirrors to detect free radicals was quickly extended during the 1930s; in this way the ethyl radical was detected in the decomposition of $Pb(C_2H_5)_4$, and the thermal decomposition of many organic molecules such as alkanes, aldehydes and ketones was shown to produce radicals.

D The production of odd-electron species

Homolysis. The experiments of Gomberg and Paneth both produced unstable free radicals by the breaking of covalent bonds. It is important to realise that there is nothing magical about covalent bonds; each requires a definite amount of energy to break it and, if that energy is available from the kinetic energy of colliding molecules, then free radicals may be formed.

If we consider a single covalent bond between two atoms A and B, then there are three ways in which bond breaking can occur:

$$A:B \longrightarrow A:^- + B^+ \qquad (1)$$

$$A:B \longrightarrow A^+ + B:^- \qquad (2)$$

$$A:B \longrightarrow A\cdot + B\cdot \qquad (3)$$

The covalent bond consists of one pair of electrons, and after fission the electrons may both reside on atom A, both on B, or one on each atom. (Other possibilities, such as the loss of one electron altogether, require much more energy, and are not caused by the collision of two molecules.) Reactions (1) and (2) both produce ions which have even numbers of electrons; this bond breaking is called heterolysis, or heterolytic fission. Reaction (3) produces two neutral species, which each have an odd number of electrons; this is called homolysis, or homolytic fission.

In the gas phase heterolytic fission is very unfavourable, as the energy required to separate the two oppositely charged ions from each other is considerable. There is no attractive force between two neutral species, however, and so in gas phase reactions the breaking of covalent bonds always leads to homolysis, and therefore free radicals.

In solutions the situation may be different. If the solvent molecules have a dipole moment then they may be strongly attracted to positive and negative ions, and if the solvent has a high relative permittivity then the energy required to separate the ions will be greatly reduced. Table 1.1 shows the dipole moments and relative permittivities of some common solvents. It is found experimentally that those towards the top of the table tend to encourage the formation of radicals, while those towards the bottom encourage the formation of ions.

Table 1.1 Properties of some common solvents

	Solvent	Relative permittivity	Dipole moment (debye)
↑	C_6H_6	2	0
Homolysis	C_6H_{12}	2	0
favoured	CCl_4	2	0
(radicals)	CS_2	3	0
	$(C_2H_5)_2O$	4	1.1
	$CHCl_3$	5	1.0
	CH_3COCH_3	21	2.9
Heterolysis	C_2H_5OH	24	1.7
favoured	CH_3OH	33	1.7
(ions)	$C_6H_5NO_2$	35	4.2
↓	H_2O	79	1.8

The two main factors which govern the breaking of covalent bonds in molecules are bond energy and temperature. High temperatures favour the formation of radicals from molecules, while low temperatures tend to produce recombination of radicals. There is a considerable range in the strength of covalent bonds: the weakest bonds require less than 50 kJ mol^{-1} to break them, while the strength of a triple bond can exceed 900 kJ mol^{-1}. In discussing these numbers it is useful to bear in mind that the average kinetic energy of molecules at room temperature is less than 10 kJ mol^{-1}.

A familiar example of a weak covalent bond is the N–N bond in N_2O_4: it has a bond energy of about 60 kJ mol^{-1}. In the liquid state the dominant species are N_2O_4 molecules, which are not odd-electron species. The liquid boils at 21 °C giving a gas which consists of about 90% N_2O_4 molecules and 10% NO_2 molecules; the NO_2 molecules are of course odd-electron species, containing 23 electrons. If the gas is warmed further, the proportion of NO_2 in the mixture increases, and at temperatures above 140 °C there is virtually no N_2O_4 present.

Rather stronger bonds are found in the halogens; the weakest bond occurs in the molecule I_2 and has a strength of about 150 kJ mol^{-1}. The halogens therefore exist entirely as diatomic molecules at room temperature. Much higher temperatures are now required to produce breaking of the covalent bond: indeed the dissociation of I_2 molecules to I atoms is not complete even at 500 °C. But as we shall see in later chapters, chemical reactions may often be started by rather small concentrations of free radicals, and heating the halogens to about 300 °C is sufficient to produce free radicals in small quantities.

Finally many of the covalent bonds which are common in organic chemistry, such as C–C, C–H, C–O and O–H, have bond energies which are in excess of 300 kJ mol^{-1}. These bonds remain unaffected by the temperatures which are encountered in the great majority of chemical reactions: they show some tendency to fission only above 500 °C. Note however that special circumstances may affect these generalisations: for example we have seen already that the C–C bond in hexaphenylethane is unusually weak, and may be broken even at room temperature.

Heating a sample is the most obvious way of providing the energy necessary to break a covalent bond, but in many cases the use of high temperatures may be undesirable or impossible. Free radicals may also be formed by the use of strong light; this allows relatively strong covalent bonds to be broken at low temperatures. We shall discuss the use of light in more detail in chapter 2, but for the moment we may note that weak bonds such as the I–I bond are broken by visible light, whereas stronger bonds such as the Cl–Cl bond require the use of ultra-violet light.

Electron transfer. Odd-electron species may also be formed from molecules or ions with complete electron shells by the removal or addition of a single electron.

The simplest method of removing an electron from an ion is to use electrolysis. For example if a solution of a salt of a carboxylic acid such as $CH_3COO^-Na^+$ is electrolysed, the ethanoate ions are attracted to the positively charged anode, where they lose an electron:

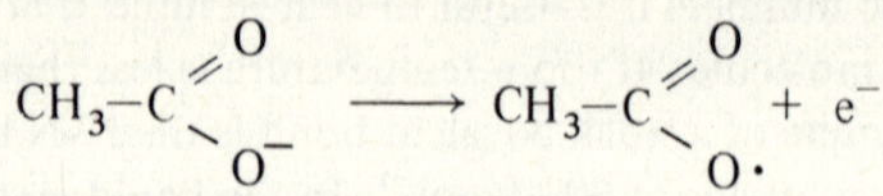

The $CH_3COO\cdot$ which is formed is an odd-electron species, and we shall study the subsequent fate of this radical in chapter 2. It seems likely that many electrolyses produce short-lived odd-electron species round the anode and cathode, often in relatively high concentrations.

Some chemical reagents are also capable of donating or removing a single electron; remember that electron-donors are reducing agents, while electron-acceptors are oxidising agents. Thus copper(I) chloride reacts with chlorine gas forming copper(II) chloride, which we may represent:

$$Cu^+ + \tfrac{1}{2}Cl_2 = Cu^{2+} + Cl^-$$

The Cu^{2+} so formed has an odd number of electrons. The ion Fe^{2+} can act as a reducing agent, and give one electron to a molecule such as hydrogen peroxide. The O–O bond is then broken, forming a hydroxide ion and an $\cdot OH$ radical:

$$Fe^{2+} + H{-}O{-}O{-}H = Fe^{3+} + HO^- + \cdot OH$$

The driving force behind these chemical methods of producing odd-electron species is either the stability of the final ion (as in Cu^{2+}) or the weakness of the original covalent bond (as in H_2O_2).

A third method of producing odd-electron species from a stable molecule is to bombard the molecule with high-energy particles; this causes ejection of an electron, leaving a positive ion. In practice a beam of electrons is normally used for the bombardment. This is a particularly vigorous method, as for many stable molecules the energy required to remove an electron (the ionisation energy) is much greater than the covalent bond energies. The initial ejection of an electron may therefore be followed by considerable disruption of the molecule, hence electron bombardment has little value as a method of initiating chemical reactions. However the energies of the bombarding electrons are easily controlled, and so this method can be used to measure ionisation energies. Electron bombardment is also important to analytical chemists as it forms the basis of mass spectrometry, as we shall see in chapter 6.

E How to spot odd-electron species and their reactions

One of the main problems encountered by those studying odd-electron species for the first time is the difficulty in distinguishing between species with odd and even numbers of electrons, and of recognising when a reaction involves odd-electron species. This section contains guidelines for the bemused, and is intended to be helpful rather than comprehensive.

Simple ions. Bear in mind the general pattern of ionic charges in the main group elements:

Group	I	II	III	IV	V	VI	VII	VIII
Charge	+1	+2	+3	+4	−3	−2	−1	0

All ions with the appropriate charge for their group have complete electron shells, and are not odd-electron species. Disregard also ions whose charges differ by two from that expected, such as Pb^{2+}. Odd-electron species have charges differing by one from those expected: simple examples are Xe^+ and Na (neutral). In the first transition series it is necessary to calculate the number of electrons from the atomic number and charge; in the case of complex ions, the use of oxidation numbers can simplify this calculation, as we shall see in chapter 5.

Covalent molecules. The general pattern for the number of covalent bonds formed by each element is:

Group	III	IV	V	VI	VII	VIII
Number of bonds	3	4	3	2	1	0

In odd-electron species one atom will have a number of bonds which differs from the general pattern by an odd number. Very conveniently, it does not matter for this purpose whether the bonds are drawn correctly or not. For example you might be unsure whether to represent NO_2 as:

$$O{=}N{=}O \qquad \text{or} \qquad O{-}N{=}O$$

but it makes no difference here; in both cases one of the atoms has the wrong number of bonds. Simple examples of odd-electron species are therefore NO, CH_3 and ClO_2, but not SO_2 or XeF_2.

If an ion contains covalent bonds, then the same rules apply as for neutral molecules, but add one bond for each minus charge on an atom, and subtract one bond for each plus charge. Charged odd-electron species include CH_4^{+} and O_2^{-}, but not NH_4^{+} or CH_3^{-}.

Reactions. There is no single clue which allows us to recognise when a reaction involves odd-electron species, but look out for the following conditions:

(a) the reaction requires strong light, visible or ultra-violet;
(b) the reaction is started by the addition of organic peroxides, or H_2O_2;
(c) the reaction is started by the addition of $Pb(CH_3)_4$;
(d) the reaction is terminated by inhibitors such as quinones;
(e) the reaction proceeds by a chain mechanism;
(f) in organic reactions, the product is symmetrical and contains twice as many carbon atoms as the starting material;
(g) the reaction is catalysed by a transition metal ion.

Each of these conditions we shall discuss in more detail later, but for the moment we may regard any one of these properties as evidence for the involvement of odd-electron species as intermediates. Look particularly for the use of light, which is a common indicator of free radicals both in examination questions and in real life.

2 The alkanes

A Introduction

The normal starting point for any course of organic chemistry is a study of the alkanes. The alkanes are hydrocarbons which contain only single covalent bonds, and they are usually thought of as being very unreactive. It is certainly true that they do not react with the great majority of reagents which are commonly used in organic chemistry: dilute acids and alkalis, metal halides, potassium permanganate and lithium aluminium hydride are all without effect. So, under the mild conditions in which organic reactions are normally conducted, we may expect alkanes and, in more complex organic molecules, simple alkyl groups to be inert. This lack of reactivity may be attributed to the high values of the C–H and C–C bond energies (435 and 346 kJ mol^{-1}), and the inability of alkanes to dissociate to form stable ions.

However under more vigorous conditions the alkanes do undergo a series of reactions which are of great theoretical and practical interest. These involve the formation of free radicals, and are called chain reactions; they differ quite markedly from other organic reactions. Their practical importance stems from the fact that alkanes are the major constituents of crude oil, one of our major sources of carbon compounds. Much of petroleum chemistry is concerned with free radical reactions.

This chapter deals with the major reactions of the alkanes: substitution, cracking and combustion. As the free radicals involved are not very selective reagents, we shall see that these reactions often give rather mixed products. Although similar reactions can be conducted with a wide variety of other organic compounds, this lack of selectivity, combined with the absence of large scale natural sources of complex molecules, means that they are seldom of use. In contrast the alkanes are widely available, and these reactions are of great practical importance.

B Chlorination of alkanes

Production of radicals. When methane and chlorine are mixed at room temperature no reaction occurs between them. This is typical of the inertness of the alkanes and, when bromine is used instead of chlorine, this forms the basis of the simple test for distinguishing between alkanes and alkenes. An alkane will not discharge the red colour of bromine, whereas an alkene readily undergoes an addition reaction forming a colourless product. However if the mixture of methane and chlorine is exposed to a bright visible or ultra-violet light, a reaction will occur: one of the hydrogen atoms in the methane molecule is replaced by a chlorine atom. This is called a substitution reaction:

$$CH_4 + Cl_2 \longrightarrow CH_3Cl + HCl$$

As the reaction proceeds only in the presence of light, it is called a photochemical reaction. More precise experiments show that the reaction occurs only if the light has a wavelength shorter than about 500 nm, that is if it falls in the violet or ultra-violet part of the spectrum. This reaction does not proceed simply by a chlorine molecule colliding with a methane molecule, but by a series of reactions of which the first is the dissociation of Cl_2 molecules into chlorine atoms. This dissociation is caused by the light. If we are to understand why light has this effect, we shall need to know a little more about the nature of light itself.

It is quite common to think of light as a sort of wave motion; this is how we come to describe light of a particular colour as having a definite wavelength and frequency. But the real nature of light is rather more complex than this would suggest, and many experiments, including the methane–chlorine reaction, can be explained only if a beam of light is thought of as a stream of particles called photons. These photons have no mass, they move in straight lines, and each one has a definite amount of energy. The energy of each photon (E) is related to the frequency of the light (ν) by the equation:

$$E = h \times \nu$$

where h is a constant, called Planck's constant. Note that this does not replace the wave theory of light which is still required to explain effects such as interference and diffraction. Instead we must think of light as being best described by a particle theory on some occasions, and by a wave theory on others; this is known as wave-particle duality. Planck's constant relates a wave property (frequency) to a particle property (photon energy).

We can now calculate the energy carried by a photon of light of wavelength 400 nm, which is about the limit of visible radiation. For any wave motion:

$$\text{speed} = \text{frequency} \times \text{wavelength}$$

So as the speed of light is 3×10^8 m s^{-1}, we have:

$$3 \times 10^8 = \text{frequency} \times 400 \times 10^{-9}$$

$$\text{Therefore frequency} = 7.5 \times 10^{14} \quad \text{Hz}$$

$$\text{As} \quad E = h \times \nu$$

$$\text{photon energy} = 6.63 \times 10^{-34} \times 7.5 \times 10^{14} \quad \text{J}$$

$$= \underline{5 \times 10^{-19}} \quad \text{J}$$

This can be compared with the relevant covalent bond energies. The bond energy for one mole of Cl_2 molecules is 242 kJ; as one mole of Cl_2 contains 6×10^{23} bonds, the energy for one Cl–Cl bond is given by:

$$E(\text{Cl–Cl}) = \frac{242 \times 10^3}{6 \times 10^{23}} \text{ J}$$
$$= \underline{4 \times 10^{-19}} \text{ J}$$

In methane, the bond energy for a C–H bond is 435 kJ mol^{-1}, and so:

$$E(\text{C–H}) = \frac{435 \times 10^3}{6 \times 10^{23}} \text{ J}$$
$$= \underline{7.2 \times 10^{-19}} \text{ J}$$

It is therefore clear that a single photon of light of wavelength 400 nm has enough energy to break a Cl–Cl bond, but not a C–H bond in methane. The relationships between wavelength (λ), frequency (ν) and photon energy (E) are shown diagrammatically in figure 2.1.

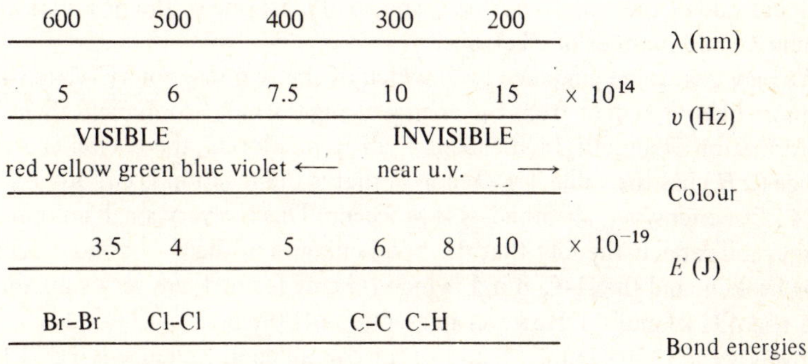

2.1 The electromagnetic spectrum

If we assume that a chlorine molecule may absorb only one photon at a time, we can see from figure 2.1 that light with wavelength greater than 500 nm has insufficient energy to break a Cl–Cl bond. As such light is also incapable of initiating the methane–chlorine reaction, this confirms that the first step of that reaction is the formation of chlorine atoms:

$$Cl_2 \xrightarrow{\text{light}} Cl\cdot + Cl\cdot \qquad (1)$$

In passing we may note that there is no possibility of ions being formed; the energy required to form a Cl^+ ion and a Cl^- ion from a chlorine molecule is very much higher, about 19×10^{-19} J.

As we shall see later, there are other less convenient methods of initiating the methane–chlorine reaction, and they also suggest that the first step is the formation of chlorine atoms.

Propagation. The chlorine atoms which are formed when light is passed through the reaction mixture are very reactive species. They are capable of reacting with many of the molecules with which they happen to collide; in this case with chlorine molecules and methane molecules. The concentration of chlorine atoms is usually very low so that, for the moment, we may neglect collisions between pairs of chlorine atoms.

If a chlorine atom collides with a chlorine molecule a chemical reaction may occur:

$$Cl\cdot + Cl_2 \longrightarrow Cl_2 + Cl\cdot \quad (2)$$

but as this reaction does not produce any overall chemical change, it will be of no significance to us. If a chlorine atom collides with a methane molecule, then there are two possible reactions:

$$Cl\cdot + CH_4 \longrightarrow CH_3\cdot + HCl \quad (3)$$

and $$Cl\cdot + CH_4 \longrightarrow CH_3Cl + H\cdot \quad (4)$$

Notice that both of these reactions result in the formation of a new free radical, $CH_3\cdot$ or $H\cdot$; this is inevitable as the total number of electrons on the left-hand side of the equation is odd, and so at least one of the products must contain an odd number of electrons.

We may gain some guidance as to which of the two reactions (3) and (4) is the more likely to occur from the energy changes which accompany them. For the formation of a methyl radical and hydrogen chloride, the overall energy change (ΔH) has the value $+4\,\text{kJ mol}^{-1}$; that is the reaction is endothermic, and 4 kJ of energy are absorbed as it proceeds. This is a very small amount of energy, and reflects the fact that the bond energies of the C–H bond, which is being broken, and the H–Cl bond, which is being formed, are very similar (435 and 431 kJ mol^{-1}). However in reaction (4) the new C–Cl bond is relatively weak (350 kJ mol^{-1}), and so the overall energy change is $+85\,\text{kJ mol}^{-1}$. We normally expect the less endothermic of two reactions to predominate, and so in this case, when a chlorine atom reacts with a methane molecule, reaction (3) occurs and a methyl radical and hydrogen chloride are formed.

Although it is quite common to use arguments based on energy changes to predict which of two or more possible reactions will occur, we might note in passing that this procedure is not very rigorous. The rates of chemical reactions do not depend on overall energy changes, but rather on activation energies. Figure 2.2 shows the way in which the energy varies during a typical reaction; this reaction is slightly exothermic, but the general shape is unchanged for an endothermic reaction. As the reaction proceeds, the energy first rises as bond-breaking begins; it reaches its maximum value when the old covalent bond is partly broken and the new bond is partly made. The energy then falls again as the new bond is completed and the old bond is completely destroyed. The activation energy of the reaction (E_a) is the difference in energy between the initial reactants and the point of highest energy; the lower the activation energy, the higher the reaction rate. When two reactions are competing with each other the faster one predominates.

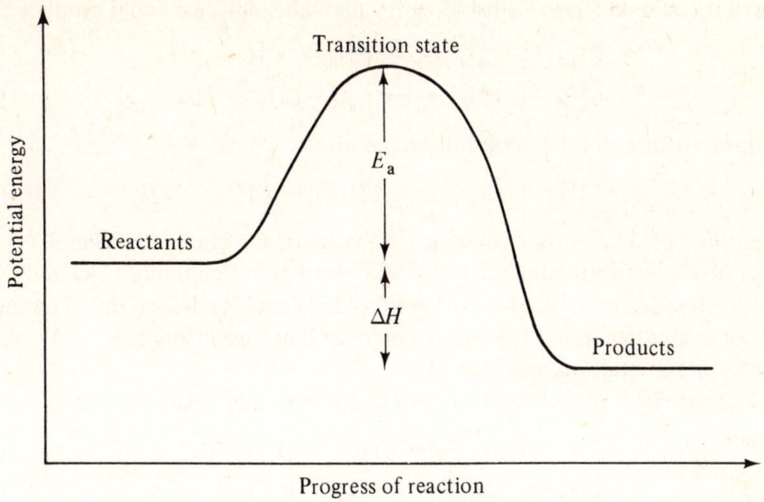

2.2 *Variation of energy during a reaction*

There is no general relationship between the activation energy (E_a) and the overall energy change (ΔH), although E_a must always be the greater (ie more endothermic). Nevertheless in most comparable cases it is found that the values of ΔH tend to fall in the same order as those of E_a, and thus energy changes may be used to predict the course of a reaction. Figure 2.3 compares the variation of energy as reactions (3) and (4) proceed.

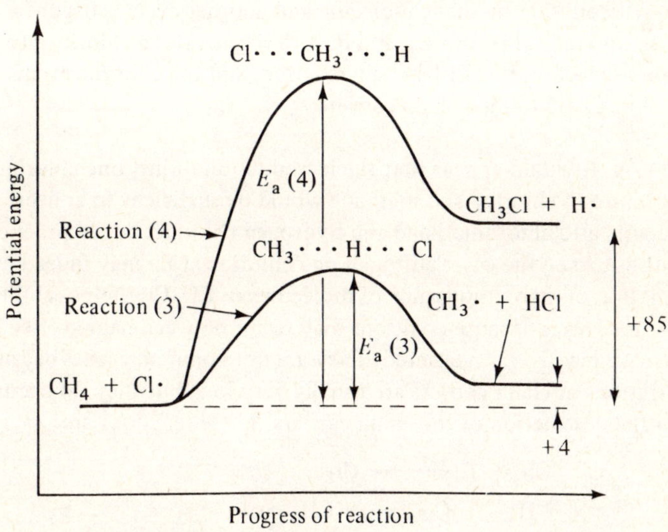

2.3 *The reaction of CH_4 and Cl*

In reaction (3) a methyl radical was formed, along with hydrogen chloride; $CH_3\cdot$ is also a reactive species, just as $Cl\cdot$ is, and we can envisage three reactions which it could undergo. Collision with a methane molecule could produce:

$$CH_3\cdot + CH_4 \longrightarrow CH_4 + CH_3\cdot \qquad (5)$$

or

$$CH_3\cdot + CH_4 \longrightarrow CH_3{-}CH_3 + H\cdot \qquad (6)$$

Collision with a chlorine molecule would give:

$$CH_3\cdot + Cl_2 \longrightarrow CH_3Cl + Cl\cdot \qquad (7)$$

The values of ΔH for these three reactions are 0, +89 and −102 kJ mol^{-1} respectively, reflecting the relative weakness of a Cl–Cl bond (242 kJ mol^{-1}), and the relative strength of a C–H bond (435 kJ mol^{-1}). Reaction (7) therefore predominates; reaction (5) produces no overall change in any case, and reaction (6) is strongly endothermic.

The net result of the two reactions which have now taken place is:

$$Cl\cdot + CH_4 \longrightarrow CH_3\cdot + HCl \qquad (3)$$

$$\underline{CH_3\cdot + Cl_2 \longrightarrow CH_3Cl + Cl\cdot} \qquad (7)$$

$$CH_4 + Cl_2 \longrightarrow CH_3Cl + HCl$$

As these two steps together convert the reactants methane and chlorine into the products chloromethane and hydrogen chloride, they are called the *propagation* steps.

As we have seen, neither reaction (3) nor reaction (7) actually consumes free radicals; the chlorine atom that is used up in reaction (3) is regenerated in reaction (7). This chlorine atom is then capable of undergoing further reaction with another methane molecule and another cycle is begun: a second methyl radical is formed, and in turn regenerates a chlorine atom. This type of reaction is called a chain reaction, and the chlorine atoms and methyl radicals are termed chain carriers.

Termination. It would appear that the introduction of just one chlorine atom into a mixture of chlorine and methane would be sufficient to convert all of the reactants into chloromethane and hydrogen chloride, but in practice this turns out not to be the case: although one chlorine atom may indeed cause the formation of many thousands of molecules of CH_3Cl, the possibilities are not infinite. This is because collisions may occur between pairs of free radicals, and these we have so far neglected. They are not common events because the concentrations of chain carriers are usually very low, but they do occur and result in the destruction of the chain carriers. These reactions are:

$$Cl\cdot + Cl\cdot \longrightarrow Cl_2 \qquad (8)$$

$$CH_3\cdot + Cl\cdot \longrightarrow CH_3{-}Cl \qquad (9)$$

$$CH_3\cdot + CH_3\cdot \longrightarrow CH_3{-}CH_3 \qquad (10)$$

As these processes end the chain reaction they are called *termination* steps. Reactions (8) and (9) merely form chlorine and chloromethane, which are present in the reaction mixture already, whereas reaction (10) produces ethane, which is not present. In practice a small amount of ethane is found among the products of the reaction of chlorine with methane.

As the collision of two radicals is a random process, some chains may be terminated very rapidly whereas others may be very long indeed. We can calculate the average chain length if we measure the number of photons absorbed by the reaction mixture, and the number of moles of product formed. Measuring the amount of light absorbed by the mixture merely involves measurement of the total power output of the lamp, and also the fraction of the incident light which is actually absorbed by the chlorine molecules; much of the light passes straight through the reaction mixture and is not absorbed at all. In a typical experiment the absorption of 20 J of energy from light of wavelength 400 nm might produce 0.15 moles of chloromethane. The frequency of light of wavelength 400 nm, as we have seen on p. 12, is given by:

$$\text{frequency} = \frac{\text{speed of light}}{\text{wavelength}} = \frac{3 \times 10^{8}}{400 \times 10^{-9}} \text{ Hz}$$

$$= 7.5 \times 10^{14} \text{ Hz}$$

$$\text{Therefore energy of one photon} = h \times \nu$$

$$= 6.63 \times 10^{-34} \times 7.5 \times 10^{14} \text{ J}$$

$$= 5 \times 10^{-19} \text{ J}$$

$$\text{In 20 J, number of photons} = \frac{20}{5 \times 10^{-19}} = 4 \times 10^{19}$$

$$\text{Now in 0.15 moles of } CH_3Cl\text{, number of molecules} = 0.15 \times 6 \times 10^{23}$$

$$= 9 \times 10^{22}$$

$$\text{Therefore average chain length} = \frac{\text{number of molecules of } CH_3Cl}{\text{number of photons absorbed}}$$

$$= \frac{9 \times 10^{22}}{4 \times 10^{19}} = 2200$$

Other initiation methods. Although the production of chlorine atoms by the use of visible and ultra-violet light is the most convenient way of initiating the methane–chlorine reaction in the laboratory, there are several other methods which can be used. They are of importance because each illustrates another way in which radicals may be formed.

The energy required to break a Cl–Cl bond is 242 kJ mol^{-1}. As discussed in chapter 1, this is too strong to allow significant numbers of Cl atoms to be formed by ordinary collisions at room temperature. But as the temperature rises the fraction of molecules with the very high energy required to break a Cl-Cl bond increases rapidly, and at a temperature of about 300 °C there are sufficient chlorine atoms present for the reaction with methane to occur. Thus although the methane-chlorine reaction is a photochemical reaction at room temperature, it becomes an ordinary thermal reaction above 300 °C. Remember however that it is only a very small proportion of chlorine molecules which have enough energy to dissociate; the average molecular kinetic energy is still less than 10 kJ mol^{-1}.

The reaction may also be initiated by introducing sodium vapour into the reaction mixture; this generates Cl atoms by the reaction:

$$Na\cdot + Cl_2 \longrightarrow NaCl + Cl\cdot$$

The affinity of chlorine for sodium is greater than its affinity for another chlorine atom; sodium chloride is of course ionic, and we shall discuss the reasons for its stability in chapter 4.

Lastly the methane-chlorine reaction may be initiated by introducing lead tetramethyl, $Pb(CH_3)_4$. As we saw in chapter 1 this has very weak Pb-C bonds, and on warming dissociates into lead atoms and methyl radicals. These $CH_3\cdot$ radicals then react with chlorine molecules forming Cl atoms as in reaction (7), and the chain is begun.

Purity of the products. In inorganic chemistry it is quite common for reactions to produce 100% yield: the precipitation of AgCl on mixing NaCl and $AgNO_3$ solutions is a simple example. But in organic chemistry it is much less usual for reactions to give very high yields, and the separation of the desired product from other by-products, and from unreacted starting material, can be a difficult and costly business. Here industry and the laboratory chemist do not always face the same problems; industry may be better placed to use an expensive item of equipment in the knowledge that it will pay for itself with regular use, or to utilise a by-product which the laboratory chemist may simply not require. On the other hand industry is likely to be more concerned with the cost of raw materials, and the ready availability of a compound may be very important.

As free radicals are generally reactive species, they can react to give very mixed products. The reaction of methane with chlorine is a relatively simple example, but even here chloromethane is not the only substance formed. Early in the reaction a chlorine atom can react only with a methane molecule; but as the reaction progresses the concentration of CH_3Cl builds up, and it becomes more likely that a chlorine atom will react with a CH_3Cl molecule than with a CH_4 molecule. This produces:

$$Cl\cdot + CH_3Cl \longrightarrow CH_2Cl\cdot + HCl$$

The new chain carrier $CH_2Cl\cdot$ then reacts with a chlorine molecule:

$$CH_2Cl\cdot + Cl_2 \longrightarrow CH_2Cl_2 + Cl\cdot$$

Overall the reaction that has occurred is:

$$CH_3Cl + Cl_2 \longrightarrow CH_2Cl_2 + HCl$$

Further reaction can give $CHCl_3$ and CCl_4. In the laboratory this prevents the methane-chlorine reaction from being a useful way of preparing chloromethane; a route from methanol is more convenient. But industrially the chlorination of methane is very important; all the products find uses as solvents, and their separation is not difficult.

When longer chain alkanes are treated with chlorine the number of possible products is even greater. A chlorine atom abstracts a hydrogen atom from the alkane, and if there are several different types of hydrogen atom then in principle all possible reactions may occur. Thus in the monochlorination of propane, both 1- and 2-chloropropane are formed:

$$CH_3{-}CH_2{-}CH_3 + Cl_2 \longrightarrow CH_3{-}CH_2{-}CH_2{-}Cl + HCl \text{ (1-chloropropan}$$

and

$$CH_3{-}CH_2{-}CH_3 + Cl_2 \longrightarrow CH_3{-}CHCl{-}CH_3 + HCl \text{ (2-chloropropane)}$$

As there are six hydrogen atoms which are bonded to carbon atoms at the end of the carbon chain, and only two bonded to the central carbon atom, it might be imagined that 1-chloropropane would be the major product, and that only small amounts of 2-chloropropane would be formed. But it is found that a hydrogen on the central carbon atom is a little easier to remove than one on a terminal carbon atom, and in practice the amounts of 1- and 2-chloropropane are very similar. The ease with which hydrogen atoms may be abstracted depends on the relative stabilities of the alkyl radicals which are formed; in general the order of stability of radicals is:

$$\text{primary} < \text{secondary} < \text{tertiary},$$

where a primary radical is of the type $R{-}CH_2\cdot$, a secondary radical is $R{-}\dot{C}H{-}R$, a tertiary radical $R{-}\dot{C}(R){-}R$, and R is an alkyl group.

Bromination and iodination. Alkanes react with bromine under the same conditions as chlorine; the reaction is usually initiated by strong light, although heating can be used. But whereas the reaction with chlorine is vigorous and can even become explosive under suitable conditions, the reaction of bromine is rather slow. The mechanism of the bromination of methane is the same as that of chlorination, but for the propagation step:

$$Br\cdot + CH_4 \longrightarrow CH_3\cdot + HBr$$

$\Delta H = +69\ \text{kJ mol}^{-1}$, as compared with $+4\ \text{kJ mol}^{-1}$ when chlorine is used. This reflects the weakness of the H-Br bond compared with the H-Cl bond;

this step is therefore much slower. Furthermore the termination steps become relatively more important, and the average chain length is reduced.

The reaction with iodine is slower still; in fact the reaction is reversible, and it is more usual to use it as a method of preparing alkanes from iodoalkanes rather than the other way round.

C Cracking

Crude oil. Undoubtedly the most important natural source of organic compounds of our time is crude oil. The annual turnover of the oil industry is measured in thousands of billions of pounds, and the political importance of oil can hardly be overestimated. Oil has been formed by the slow decay of organic matter, which has been trapped by the deposition of impervious layers of rock above it. The formation of oil is roughly similar to the formation of methane gas in stagnant ponds.

Although the precise composition of crude oil depends on where it is found, all crude oil is composed mostly of a mixture of hydrocarbons, with alkanes forming by far the largest group of compounds. Crude oil is first separated into a series of different mixtures by fractional distillation; table 2.1 shows the composition and boiling ranges of the main fractions.

Table 2.1 *Main fractions of crude oil*

Fraction	*Number of carbon atoms*	*Boiling range (°C)*	*Uses*
Gas	1–4	up to 20	Fuel
Petroleum ether	5–7	20–80	Solvent
Gasoline	6–11	60–180	Motor fuel
Kerosine	10–13	180–250	Lighting
Gas oil	13–18	200–340	Diesel oil, heating
Lubricating oil	18–20	>340	Lubricants
Fuel oil	20–30	>340	Heating
Bitumen	30–	>340	Tar

The figures given here can only be approximate; remember that there is only one alkane containing two carbon atoms, whereas $C_{15}H_{32}$ has many isomers, all with slightly different properties. Note that there is no attempt to separate individual alkanes from crude oil at this stage; this would be extremely difficult and, as we shall see, not at all useful.

Cracking. We have already seen that the C–C and C–H bonds which are found in the alkanes are extremely strong; their bond energies fall in the range 350–450 kJ mol^{-1}. They therefore remain unchanged in the great majority of chemical reactions with which we are familiar. But, if sufficient heat is applied to them, their bonds can be broken, and it is found experimentally that a temperature in excess of 500-600 °C is usually required. This may be compared with a temperature of under 100 °C to break a Pb-C bond in lead tetramethyl, and of about 300 °C to break a Cl-Cl bond. When alkanes are heated in this way they decompose to form hydrocarbons with shorter chains; the process is called cracking. Because of this, the higher fractions of crude oil are not usually distilled at atmospheric pressure (where they would decompose), so instead they have to be vacuum distilled.

If we consider octane, C_8H_{18}, as an example of an alkane, we may appreciate some of the complexities of the reaction. Octane contains seven C–C bonds, all with very similar bond energies, and the initial bond-breaking may be:

$$\begin{aligned} & C_8H_{18} \longrightarrow CH_3\cdot + \cdot CH_2{-}CH_2{-}CH_2{-}CH_2{-}CH_2{-}CH_2{-}CH_3 \\ \text{or} \quad & \longrightarrow CH_3{-}CH_2\cdot + \cdot CH_2{-}CH_2{-}CH_2{-}CH_2{-}CH_2{-}CH_3 \\ \text{or} \quad & \longrightarrow CH_3{-}CH_2{-}CH_2\cdot + \cdot CH_2{-}CH_2{-}CH_2{-}CH_2{-}CH_3 \\ \text{or} \quad & \longrightarrow CH_3{-}CH_2{-}CH_2{-}CH_2\cdot + \cdot CH_2{-}CH_2{-}CH_2{-}CH_3 \end{aligned} \quad (11)$$

Here the choice of a symmetrical alkane has simplified matters; the isomeric 2-methylheptane, $(CH_3)_2\ CH{-}CH_2{-}CH_2{-}CH_2{-}CH_2{-}CH_3$, would have produced even more radicals.

These alkyl radicals may undergo many different reactions. A radical may abstract a hydrogen atom from a neutral molecule:

$$CH_3{-}CH_2\cdot + C_8H_{18} \longrightarrow CH_3{-}CH_3 + C_8H_{17}\cdot \quad (12)$$

In this case an alkane with a shorter chain than the original is formed. Note that the alternative reaction in which a longer chain alkane is formed is highly endothermic (see p. 16), and is therefore unlikely to occur:

$$CH_3{-}CH_2\cdot + C_8H_{18} \longrightarrow CH_3{-}CH_2{-}C_8H_{17} + H\cdot$$

Alternatively a radical may simply decompose on its own, producing an alkene and another radical:

eg $C_3H_7\cdot \longrightarrow C_2H_4 + CH_3\cdot$

These reactions are both propagation steps, as neither involves a change in the number of free radicals.

Two free radicals may also react with each other; when they do, it seems that the dominant reaction is a hydrogen atom transfer reaction:

$$\begin{aligned} \text{eg} \quad & CH_3{-}CH_2\cdot + \cdot CH_2{-}CH_2{-}CH_3 \longrightarrow CH_2{=}CH_2 + CH_3{-}CH_2{-}CH_3 \\ \text{or} \quad & \longrightarrow CH_3{-}CH_3 + CH_2{=}CH{-}CH_3 \end{aligned}$$

Here an alkane and an alkene are formed. There may also be some simple recombination:

eg $CH_3\cdot + \cdot CH_2{-}CH_3 \longrightarrow CH_3{-}CH_2{-}CH_3$

In both these reactions radicals are destroyed, and they are therefore termination reactions.

Clearly as the reaction proceeds there is a build-up of short chain alkanes and alkenes, and these may then undergo further fission as in reaction (11), or react with other free radicals as in reaction (12). Note that it is not possible in principle to crack an alkane simply into two shorter alkanes, as there are insufficient hydrogen atoms:

eg $C_4H_{10} \not\longrightarrow C_2H_6 + C_2H_6$ or $CH_4 + C_3H_8$

but $C_4H_{10} \longrightarrow C_2H_4 + C_2H_6$ or $CH_4 + C_3H_6$

so alkenes are inevitably formed in cracking.

The result of the cracking reaction is therefore the production of a complex mixture of short chain alkanes and alkenes. This is scarcely the sort of reaction which is beloved of laboratory chemists; the temperature required is inconveniently high, and the yield of any one substance is very low. But it is much more important in the petrochemical industry; high temperatures are easily attained, and in practice pure octane would not be cracked alone. A mixture of alkanes obtained from fractional distillation is used, and the fact that the products are highly mixed is much less of a disadvantage.

Uses of cracking. Although we have now seen that the thermal cracking of alkanes is feasible industrially, we have not yet answered the question why it might be desirable. There are two main reasons why cracking is conducted on an enormous scale. Firstly much the largest demand for hydrocarbons is for use as fuels, but unfortunately the pattern of demand does not match that of availability from fractional distillation. Thermal cracking conveniently converts relatively involatile alkanes, for which supply outstrips demand, into more flammable alkanes, for which the reverse is true.

The second advantage of thermal cracking derives from the fact that it produces alkenes. As well as being used as a fuel, crude oil is a major source of organic chemicals. Unfortunately as the alkanes are rather unreactive, they are not a very useful starting point in the manufacture of new materials. But the alkenes react with a wide variety of reagents under mild conditions, and are important precursors of organic compounds. Propene and ethene are formed in the largest quantities in thermal cracking. Propene is used in the production of polymers such as polypropene, polyurethanes and epoxyresins, and in the formation of propanone and phenol; ethene is used as a starting material for polythene, PVC (polyvinylchloride) and polystyrene, for ethylene glycol and polyesters, for ethanol and for lead tetraethyl.

D Combustion

Combustion in air is the one reaction of which virtually all organic compounds are capable. The combustion of alkanes is the most important example, because of the large natural resources of crude oil, but it often receives rather scant attention in text-books. The reaction is of relevance to us because it involves free radicals.

Alkanes do not react when they are mixed with air or oxygen at room temperature; the reaction requires the application of a flame or a spark to initiate it. In the presence of excess oxygen the final products are carbon dioxide and water, and of course the reaction is highly exothermic:

$$\text{eg} \quad C_3H_8 + 5O_2 \longrightarrow 3CO_2 + 4H_2O \quad \Delta H = -2220\,\text{kJ mol}^{-1}$$

Under suitable conditions the reaction is explosive (we might note here that a flame is really a controlled explosion, in which there is a continuous supply of reactants, and from which the products may escape); the explosion of alkanes and oxygen is discussed on p. 24. If the supply of oxygen is limited carbon monoxide or carbon may be formed. If the temperature is kept low it is possible to avoid an explosion, and a wide variety of oxidation products such as alcohols, aldehydes and ketones can be obtained.

The spark initiates the reaction by producing alkyl radicals:

$$R{-}H \longrightarrow R\cdot + H\cdot$$

These radicals then react readily with oxygen molecules, forming alkylperoxy radicals:

$$R\cdot + O_2 \longrightarrow R{-}O{-}O\cdot$$

These radicals are rather unreactive, and react with each other:

$$R{-}O{-}O\cdot + R{-}O{-}O\cdot \longrightarrow R{-}O\cdot + R{-}O\cdot + O_2$$

These more reactive products can then undergo a variety of reactions, depending on the exact conditions: for example hydrogen abstraction produces an alcohol:

$$R{-}O\cdot + R{-}H \longrightarrow R{-}O{-}H + R\cdot$$

Knocking. The most important example of the combustion of alkanes is the burning of petrol in the internal combustion engine. Unfortunately the fuel obtained by fractional distillation of crude oil does not burn very well in spark-ignition engines; it tends to cause 'knocking', or premature detonation. Ideally the petrol–air mixture should burn rapidly but smoothly on sparking; when knocking occurs, unburnt fuel suddenly explodes in front of the flame, which reduces power and causes engine wear.

It is found that knocking is increased when the fuel consists largely of straight-chain alkanes, and is reduced when alkanes with highly branched chains are used. The tendency of a fuel to knock is measured by its octane number. Heptane, a straight-chain alkane, is assigned an octane number of 0, and 2,2,4-trimethylpentane, often wrongly called iso-octane, an octane number of 100. Fuels are compared for their knocking properties with mixtures of heptane and 2,2,4-trimethylpentane; the octane number is then the percentage of 2,2,4-trimethylpentane in the mixture with equivalent burning properties.

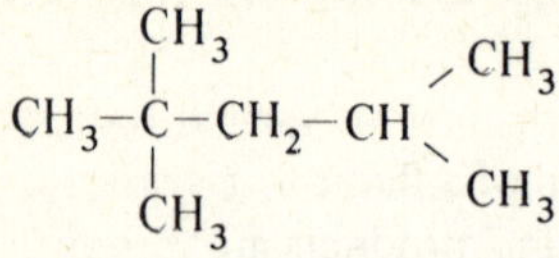

2, 2, 4-trimethylpentane ('iso-octane')

Methods have therefore been developed for converting straight-chain alkanes into their branched-chain isomers; $AlCl_3$ can be used as a catalyst for this reaction, but this is an example of a reaction involving ions and not radicals.

Diesel engines are the opposite of spark-ignition engines; here the reaction is initiated by compression and high temperatures rather than by a spark, and diesel engines run more smoothly on straight-chain alkanes.

Explosions. We have already encountered two examples of explosions in alkane chemistry; methane–chlorine mixtures explode in bright light, and methane–oxygen mixtures explode on sparking. An explosion occurs when the rate of a chemical reaction increases very rapidly, so that all the reactants are consumed in a very short period of time. These reactions are exothermic, so an explosion produces a sudden increase in temperature and pressure.

There are essentially two ways in which an explosion may occur. It is well known that the rates at which chemical reactions occur increase rapidly with temperature; in rough terms, reaction rates double when the temperature is increased by 10 °C. Most reactions produce heat as they proceed, but this is normally dissipated to the surroundings, and only a modest temperature increase is observed. But if the heat of reaction is generated faster than it can be dissipated there is a large temperature rise, which causes the reaction to proceed faster. This in turn produces more heat, giving another increase in reaction rate, and so the temperature increases very rapidly. This is called a thermal explosion, and the methane–chlorine reaction is an example of this type of explosion.

The second type of explosion is called a branched-chain explosion, and the reaction of methane with oxygen is an example of this type. As we have seen, the first step is:

$$R\cdot + O_2 \longrightarrow R{-}O{-}O\cdot$$

At low temperatures the dominant reaction then leads to slow oxidation of the alkane; but at higher temperatures a hydrogen abstraction can occur:

$$R{-}O{-}O\cdot + R{-}H \longrightarrow R{-}O{-}O{-}H + R\cdot$$

The alkyl radical then combines with more oxygen, which completes the usual propagation steps, with R· and R-O-O· acting as chain carriers. However the product R-O-O-H has a very weak O-O bond (see chapter 3), and if the temperature is high enough it dissociates:

$$R{-}O{-}O{-}H \longrightarrow R{-}O\cdot + \cdot O{-}H$$

This is an important step as R-O· and H-O· are also reactive radicals and may act as chain carriers. We need not be concerned with the wide variety of radical reactions which may now occur, finally producing carbon dioxide and water; we need note only that the propagation steps produce a net increase in the number of chain carriers. Each time the reaction goes round the cycle once, the number of radicals increases, and the chain is said to branch. This increases the rate of reaction, which produces more radicals, and so the reaction goes yet faster, eventually becoming an explosion.

In fact there is a close similarity between this sort of branched-chain reaction and an atomic bomb. In the latter a neutron causes the fission of one atom, and that fission produces several more neutrons. Each of these causes the fission of another atom, producing further neutrons, and if the conditions are properly controlled the number of neutrons increases very rapidly, as does the rate of atomic fission. As each fission is a highly exothermic process, an enormous amount of energy is liberated very rapidly and the bomb explodes.

E The Kolbe synthesis of alkanes

In the reactions of alkyl radicals that we have studied so far, the initial propagation steps have always involved the fission of a covalent bond, forming two neutral radicals. Thus, in chlorination, Cl· atoms were formed by the fission of a Cl-Cl bond and, in cracking, alkyl radicals were formed by fission of a C-C bond in an alkane. This type of reaction is called homolysis, as the electron pair which originally constituted the covalent bond is split, with one electron going to each fragment.

Although homolysis is much the commonest way of forming unstable radicals such as $CH_3\cdot$, it is not the only possible method. In Kolbe's synthesis of alkanes, an electron-transfer reaction takes place; a negative ion loses one electron, thereby becoming neutral and acquiring an odd number of electrons.

In Kolbe's reaction, alkanes are formed when solutions containing the sodium salts of carboxylic acids are electrolysed. Overall the reaction which occurs is:

$$2R{-}CO_2^- + 2H_2O \longrightarrow R{-}R + 2CO_2 + 2OH^- + H_2$$

This reaction may be divided into two half-reactions, one of which occurs at the anode, and the other at the cathode. The negative carboxylate ions are attracted to the anode where they lose an electron:

$$2R{-}CO_2^- \longrightarrow R{-}R + 2CO_2 + 2e^-$$

The electrons produced in this step then go round the circuit to the cathode where the other half-reaction takes place:

$$2H_2O + 2e^- \longrightarrow 2OH^- + H_2$$

The yield of alkane is about 70%, but it is found that traces of other products are formed, including alkenes and esters.

The first step in this reaction is the loss of an electron by the carboxylate ion to the anode:

$$R{-}C{\overset{\displaystyle O}{\diagup\!\!\!\!\diagup}}\!\!\underset{\displaystyle O^-}{\diagdown} \longrightarrow R{-}C{\overset{\displaystyle O}{\diagup\!\!\!\!\diagup}}\!\!\underset{\displaystyle O\cdot}{\diagdown} + e^-$$

The resulting radical $RCO_2\cdot$ is unstable, and undergoes decarboxylation (loss of CO_2):

$$R{-}C{\overset{\displaystyle O}{\diagup\!\!\!\!\diagup}}\!\!\underset{\displaystyle O\cdot}{\diagdown} \longrightarrow R\cdot + O{=}C{=}O$$

As the R· radicals are formed in quite high concentrations around the anode, the major termination step is:

$$R\cdot + R\cdot \longrightarrow R{-}R$$

However other fates are possible for the alkyl radicals, and they account for the small amounts of other products formed. Two alkyl radicals may undergo a hydrogen atom transfer as a termination step:

eg $$C_2H_5\cdot + C_2H_5\cdot \longrightarrow C_2H_4 + C_2H_6$$

thus forming an alkane and an alkene. Alternatively an alkyl radical may collide with an R-$CO_2\cdot$ radical before it has had time to decarboxylate, thus yielding an ester:

$$R{-}C{\overset{\displaystyle O}{\diagup\!\!\!\!\diagup}}\!\!\underset{\displaystyle O\cdot}{\diagdown} + R\cdot \longrightarrow R{-}C{\overset{\displaystyle O}{\diagup\!\!\!\!\diagup}}\!\!\underset{\displaystyle O{-}R}{\diagdown}$$

The Kolbe reaction may be extended to unsymmetrical alkanes by using a mixture of the salts of two different carboxylic acids. If the ions R_1COO^- and R_2COO^- are used, then radicals $R_1\cdot$ and $R_2\cdot$ are formed, and they combine forming $R_1{-}R_1$, $R_1{-}R_2$ and $R_2{-}R_2$. In favourable cases, this mixture of products may be easily separated.

3 Unsaturated hydrocarbons

A Introduction

Unsaturated hydrocarbons, those containing carbon-carbon multiple bonds, are much more reactive than the alkanes which we studied in the last chapter. They undergo similar reactions to the alkanes under vigorous conditions - for example they burn in air giving carbon dioxide and water - but they also react with a wide variety of compounds under mild conditions. The alkenes and alkynes give addition products readily, and so if ethene is shaken with bromine, we get the following reaction:

$$CH_2{=}CH_2 + Br_2 = BrCH_2{-}CH_2Br$$

Notice that this is a particular feature of alkenes and alkynes, caused by the relative weakness of carbon-carbon multiple bonds, and is not a general feature of all elements; thus $H_2C{=}O$ does not react with bromine giving $CH_2Br{-}OBr$. In aromatic compounds such as benzene the multiple bonding is rather stronger than in ethene, and substitution reactions are very important, although addition reactions can occur.

The precise ways in which unsaturated hydrocarbons react have been studied in considerable detail. It is now known that there are broadly two different mechanisms which can apply: a reaction may involve ions, in which case the electrons remain paired up throughout, or alternatively it may involve free radicals, in which case unpaired electrons are involved at some stage. Whether a given reaction proceeds via an ionic or radical mechanism depends on the exact reaction conditions.

The question of whether a reaction proceeds via an ionic or free radical mechanism may seem rather obscure. But as we shall see, it is of more than merely academic interest as in many cases the mechanism of the reaction determines the nature of the products. Table 3.1 summarises the main products of the three reactions discussed in this chapter; in each case a free radical

Table 3.1 *Effect of reaction mechanism on products*

Reactants	*'Ionic' product*	*'Radical' product*
propene + HBr	2-bromopropane	1-bromopropane
benzene + Cl_2	C_6H_5Cl	$C_6H_6Cl_6$
polymerisation of ethene	high density polythene	low density polythene

mechanism gives a different product from an ionic mechanism. Only when we understand the factors which govern the reaction mechanism can we select the conditions which will give the highest yield of any desired product.

B Polymerisation

In industrial terms one of the most important reactions of the alkenes is polymerisation. Names such as polythene, polystyrene, PVC, PTFE (polytetrafluoroethylene) and polypropylene have become an established part of the language, and plastics have come to play an increasingly important role in our lives.

The polymerisation of alkenes is strictly speaking an addition polymerisation. When ethene gives polythene there is only one product:

$$CH_2{=}CH_2 \;+\; CH_2{=}CH_2 \;+\; CH_2{=}CH_2 \;+\; \ldots \longrightarrow$$
$$-CH_2-CH_2-CH_2-CH_2-CH_2-CH_2-$$

The double bonds of the ethene are converted into single bonds in polythene; these are the characteristic features of addition reactions. An alternative type of polymerisation, with which we shall not be concerned, is condensation polymerisation; here there are two products, the polymer and a small molecule such as water. The formation of polyesters is a simple example of condensation polymerisation:

$$HO{-}CH_2{-}COOH + HO{-}CH_2{-}COOH + HO{-}CH_2{-}COOH + \cdots \longrightarrow$$

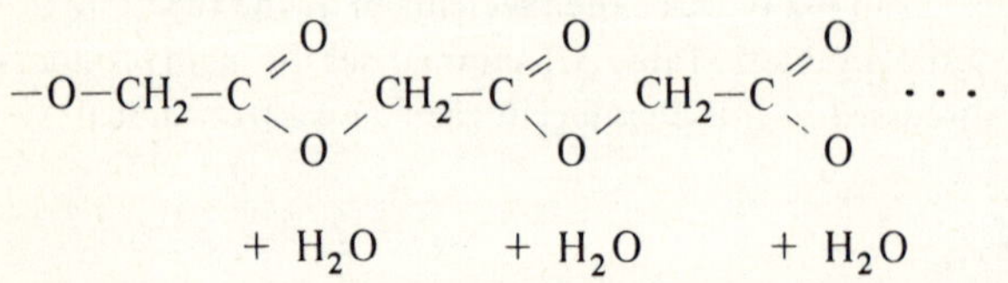

Initiation. Although the formation of polythene appears to require the presence of only one reactant, it would be quite wrong to suppose that the act of bringing together ethene molecules is sufficient to induce polymerisation. Alkenes are stable for indefinite periods if kept on their own, for although polymerisation is an exothermic reaction, it has a high activation energy, and the rate of reaction is negligible at room temperature. A number of compounds have been found to induce polymerisation even when present in very small quantities for obvious reasons they are often referred to as catalysts. This usage is not strictly correct as they do not remain chemically unchanged at the end of the reaction, which is one of the defining conditions of catalysis. We shall therefore use the term initiator, which more accurately describes their action.

Peroxides have been found to be very efficient initiators of polymerisation reactions; they contain O–O single bonds, which generally have very low bond energies. Benzoyl peroxide, C_6H_5-CO-O-O-CO-C_6H_5, is commonly used; its O–O bond energy is less than 150 kJ mol^{-1}. On warming to between 70 and 100 °C, it decomposes to form benzoyl radicals:

$$C_6H_5CO{-}O{-}O{-}COC_6H_5 \longrightarrow C_6H_5CO{-}O\cdot + \cdot O{-}COC_6H_5 \quad (1)$$

The decomposition may also be brought about by irradiation with ultra-violet light, just as in the chlorination of alkanes. Azo-compounds have also been used to initiate polymerisation:

$$R{-}N{=}N{-}R \longrightarrow R\cdot + N_2 + \cdot R \quad (2)$$

These are also well-known sources of free radicals, and in fact almost any convenient source of organic free radicals will initiate polymerisation of alkenes.

Propagation. Although oxygen atoms have rather a low affinity for other oxygen atoms, they form much stronger bonds with some other elements, particularly with carbon. The benzoyl radical formed in equation (1) is therefore rather a reactive species; its most probable fate is collision with an alkene molecule, when the following addition reaction occurs:

$$R{-}O\cdot + CH_2{=}CH_2 \longrightarrow R{-}O{-}CH_2{-}CH_2\cdot \quad (3)$$

Here the benzoyl radical has been written R-O·, and we have chosen ethene as our example of an alkene. Bond energies can give us an approximate idea of ΔH for this step:

Bonds broken		Bonds made	
C=C	+ 611 kJ mol^{-1}	C–C	– 346 kJ mol^{-1}
		C–O	– 358
			– 704

Total: $\Delta H = +611 - 704 = -93$ kJ mol^{-1}

This confirms that equation (3) is reasonable on energetic grounds.

The product R-O-CH_2-$CH_2\cdot$ is also a free radical; its most likely fate is collision with another ethene molecule, and this gives:

$$R{-}O{-}CH_2{-}CH_2\cdot + CH_2{=}CH_2 \longrightarrow R{-}O{-}CH_2{-}CH_2{-}CH_2{-}CH_2\cdot$$

This again is an exothermic step. Clearly this process will be repeated many times, and on each occasion the length of the carbon chain will increase by two atoms. The result will be a chain of considerable length, consisting almost entirely of (CH_2-CH_2) units.

Termination. If the propagation steps described above were the only possible reactions which the growing polymer could undergo, then the addition of even the smallest amount of initiator would be sufficient to polymerise any amount of alkene. In practice this is not the case, and chain termination reactions may occur, just as in the chlorination of alkanes. Because of the low concentrations of free radicals these are not very probable reactions, but they do place a limit on the length of a polymer chain.

The most obvious termination reaction is the collision of two radicals producing one longer molecule:

$$RO{-}(CH_2{-}CH_2)_n\cdot + RO{-}(CH_2{-}CH_2)_m\cdot \longrightarrow RO{-}(CH_2{-}CH_2)_{n+m}{-}OR$$

Termination can also occur if one radical abstracts a hydrogen atom from another; the products are then an alkane and an alkene:

$$RO{-}(CH_2{-}CH_2)_n\cdot + RO{-}(CH_2{-}CH_2)_m\cdot \longrightarrow RO{-}(CH_2{-}CH_2)_n{-}H + RO{-}(CH_2{-}CH_2)_{m-1}{-}CH{=}CH_2$$

Clearly termination does not occur after a set number of propagation steps; in any polymer there will be a distribution of molecules with varying chain lengths. The average chain length may be controlled by varying the concentration of the initiator; if the initiator concentration is high, then the number of growing chains will be high, and the probability of termination will be high.

Measurements of relative molecular masses, commonly based on viscosity, osmotic pressure or light scattering, allow average chain lengths to be determined, and it is found that they fall in the region 10^2–10^5. Although the average relative molecular mass is an important factor in determining the properties of a polymer, it does not describe the product completely, and the distribution of relative molecular masses is also of some importance. Note that the nature of the initiator is of no importance in determining the properties of a polymer; even for the shortest chains the initiator forms only a tiny proportion of the whole.

Substituted ethenes. We have concentrated so far on the polymerisation of ethene, the simplest alkene. But polymerisation is not restricted to ethene; a wide variety of alkenes have been polymerised, yielding products whose

properties are all slightly different. Although polymer science has not yet reached the stage where the chemical composition of a polymer may be 'designed' to produce a particular blend of properties, there are some general trends which have become apparent. Thus the introduction of halogen atoms into a polymer produces an increase in heat resistance, and the use of PTFE ($-CF_2-CF_2-CF_2-$), in non-stick frying pans is a familiar example. Increasing the hydrocarbon content tends to decrease the moisture absorption of a polymer, while increasing the chain length tends to decrease solubility.

If an alkene has two hydrogen atoms attached to one of the carbons in the double bond, ie is of the type $CH_2{=}CXY$, then polymerisation is usually rapid and irreversible. There would appear to be two ways in which the propagation step could proceed:

$$RO{-}CH_2{-}CXY\cdot + CH_2{=}CXY \longrightarrow RO{-}CH_2{-}CXY{-}CH_2{-}CXY\cdot$$

$$\text{or}\quad RO{-}CH_2{-}CXY\cdot + CXY{=}CH_2 \longrightarrow RO{-}CH_2{-}CXY{-}CXY{-}CH_2\cdot$$

In practice it is found that the former almost always predominates, and successive monomer units add head-to-tail. This is probably because the bulk of the groups X and Y makes radical attack at their end of the molecule less favourable; this is called steric hindrance. Thus the result of polymerising methylpropene, $(CH_3)_2C{=}CH_2$, is:

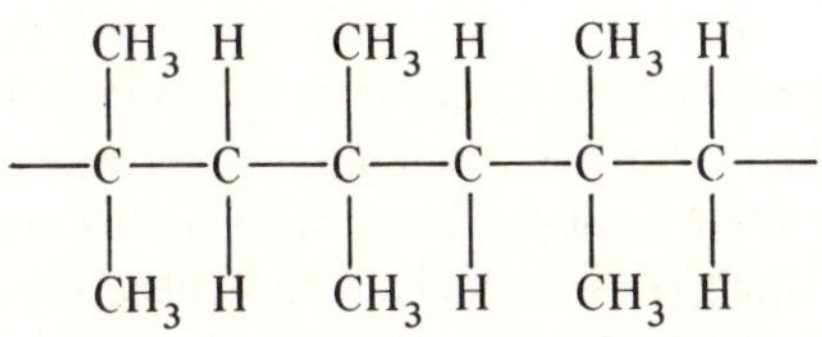

Ionic and free radical polymerisation. The examples that we have studied so far have all been free radical polymerisations. But polymerisations can also be initiated by ionic reagents, and there are some important differences between the products obtained in the two cases.

Polymers may be divided broadly into two kinds, crystalline and amorphous. This is not an entirely satisfactory distinction, because a given sample may contain both crystalline and amorphous areas, but it draws attention to the way in which polymer molecules pack together. If the polymer chains are uniform and lie closely parallel to one another, the van der Waals' forces between them are relatively strong, and a regular crystalline material results. If the polymer chains are unable to lie closely parallel, because of the irregular presence of bulky groups or side-chains, then an amorphous material is more likely.

Ethene may be polymerised commercially by heating it at high pressure in the presence of an initiator such as a peroxide, or oxygen. This is a free radical polymerisation, and the resulting product, low density polythene, has a branched-chain structure, and relatively low crystallinity. Chain branching can occur in several ways: for example a growing polymer chain can 'bite back' on itself, abstracting a hydrogen atom and creating a side-chain as the polymer reacts with further ethene molecules:

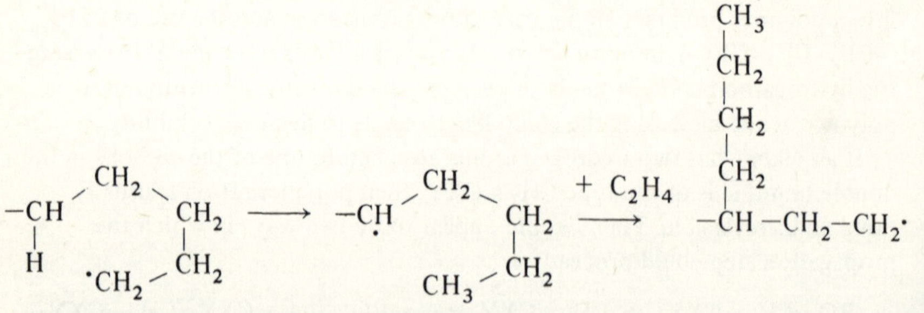

Alternatively a growing chain may abstract a hydrogen atom from an entirely different chain; this will again produce a side-chain, and can even result in the cross linking of two different polymer chains:

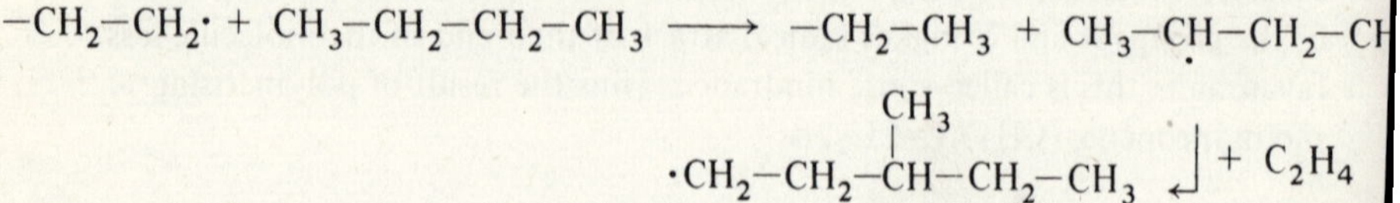

Catalysts introduced by Ziegler and Natta are also capable of inducing the polymerisation of ethene, but the reaction now proceeds via an ionic mechanism. Hydrogen abstraction is much less probable in this case, and the resulting polythene consists largely of straight-chain molecules. These molecules can pack together better, and so this high density polythene has a greater crystallinity, melting point and tensile strength. The properties of high and low density polythene are compared in table 3.2.

Table 3.2 *Comparison of high and low density polythene*

	Density $(g\ cm^{-3})$	*Crystallinity*	*Softening temperature* (°C)	*Tensile strength* $(N\ m^{-2})$
High (ionic)	0.95	95%	130	3×10^7
Low (radical)	0.92	65%	100	1×10^7

A further difference between ionic and radical polymerisation becomes apparent when an asymmetric molecule such as propene is polymerised. The Ziegler-Natta catalysts are highly stereospecific, and the relative orientation of each monomer is the same. The resulting polypropene has the structure:

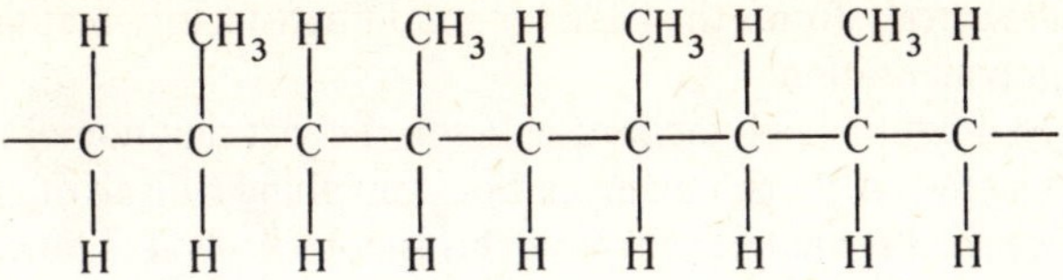

This structure is called isotactic. (Remember that the bonds round each carbon atom are really tetrahedrally arranged; we show them at right-angles here just for clarity.) In contrast radical polymerisation tends to produce more random orientations of monomer molecules; the polymer is then said to be atactic:

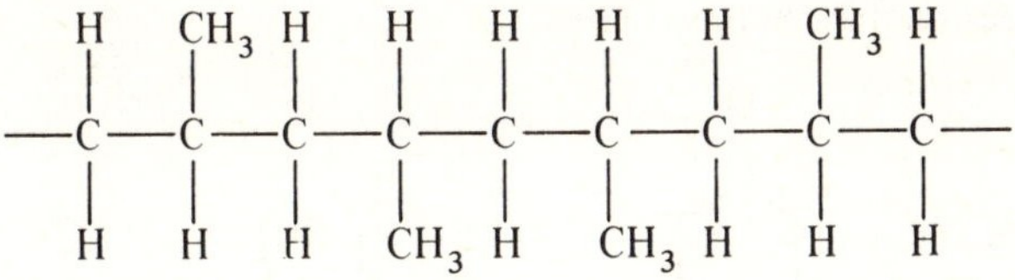

Not surprisingly the isotactic polymer chains are able to pack rather more closely than atactic chains, and so the ionic mechanism again produces polymers which are more crystalline, harder, and have a higher melting point.

Another important development in polymer science has been the introduction of copolymerisation. Here a mixture of two alkenes is polymerised, and the aim is to produce a polymer chain which contains both monomers; this is only possible if a radical initiator is used. In some cases it is found that the resulting polymer has monomer units distributed at random down the chains, whereas in a few cases it has been possible to produce a polymer in which one monomer is invariably followed by the other. There are two major merits of copolymerisation: firstly it is often possible to use alkenes which will not polymerise on their own; secondly the properties of a copolymer are usually some sort of average of the properties of the two single polymers, and in many cases it is possible to combine the merits of two polymers while losing the demerits.

Inhibitors. We have already referred to the fact that a wide variety of free radicals are capable of initiating the polymerisation of alkenes. This presents difficulties if alkenes are to be stored for any period of time; small traces of impurities could induce the polymerisation of a large stock of alkene, particularly if that stock were exposed to sunlight.

The problem is to find a method of destroying free radicals before they have time to react with an alkene molecule, starting a polymer chain. The reaction of a radical with a molecule must always produce a new radical, as the total number of electrons remains odd throughout; the only way in which the total number of radicals in a sample may be decreased is by two radicals combining forming a new covalent bond. However the addition of more radicals scarcely forms the basis for a more satisfactory method of preventing polymerisation!

The key to inhibiting polymerisation lies in the fact that not all radicals are equally reactive. Although many carbon-containing radicals are very reactive, as we shall see in later chapters this is not a general feature of all elements; nor is it even true of all organic radicals. In chapter 1 we saw that the radical $(C_6H_5)_3C\cdot$ may be kept indefinitely as long as oxygen is excluded. Quite stable organic radicals are formed when the unpaired electron can be delocalised round one or more benzene rings; we can visualise this either in terms of the overlap of adjacent *p* orbitals, or by writing resonance canonical structures. If we consider $C_6H_5O\cdot$ as an example, the delocalisation of the unpaired electron may be represented:

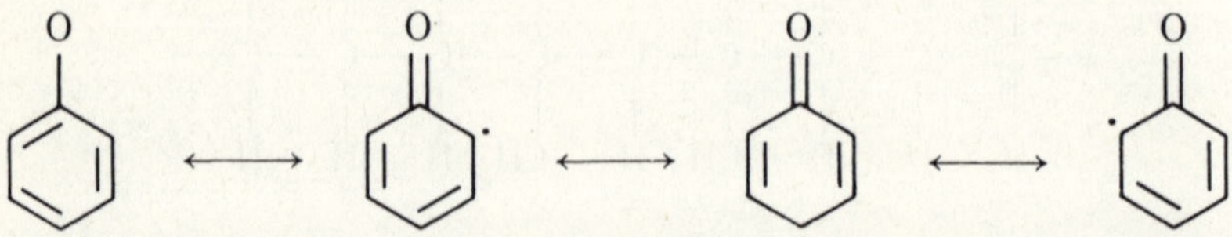

Many compounds have been shown to inhibit polymerisation: quinones, hydroquinones, aromatic amines and nitrocompounds are often used. Inhibitors work by converting a reactive radical, which would otherwise react with an alkene starting a polymer chain, into an unreactive radical, which is unable to react with an alkene. For example hydroquinone undergoes hydrogen abstraction readily, and the most likely reaction of a radical in a mixture of ethene and hydroquinone is:

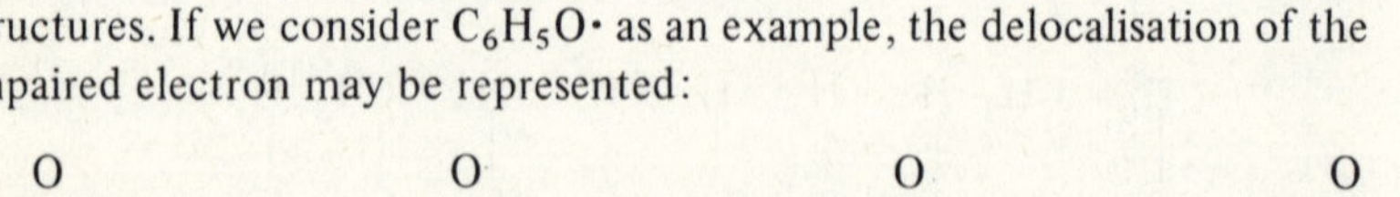

The new radical is now too unreactive to combine with ethene, and its only possible reaction is combination with another similar radical:

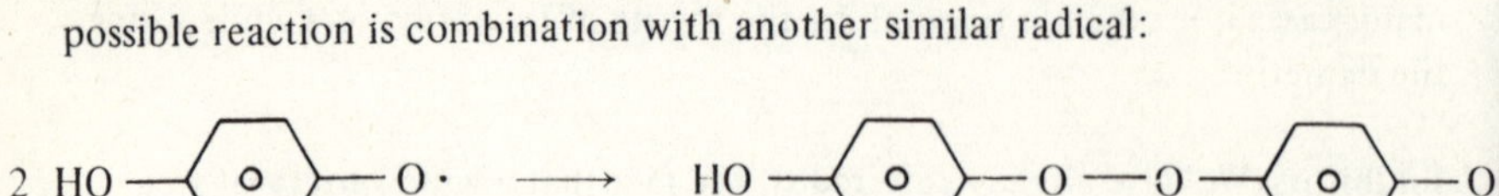

Other inhibitors react similarly: they are all capable of losing a hydrogen atom readily giving a radical which is stabilised by electron delocalisation. Inspection of the labels of most alkene bottles will reveal the presence of hydroquinone or some other inhibitor, in a concentration of between 0.1 and 1.0%.

Inhibitors are also used in the preservation of foodstuffs. Fats and oils are attacked by oxygen forming hydroperoxides in a free radical chain reaction; further oxidation then produces the aldehydes, ketones and acids which give rise to the unpleasant smell and taste of rancid food. Inhibitors act as chain-breakers in this sequence, and the presence of only small traces has a very marked effect on food storage times. Inhibitors which are used in foodstuffs are generally referred to as anti-oxidants. Similar principles apply to the use of anti-oxidants in the protection of rubbers, petrol and synthetic fibres.

C Markovnikov's rule

It has long been known that many reagents will react with alkenes giving addition reactions: halogens, hydrogen halides, acidified $KMnO_4$, sulphuric acid and more recently B_2H_6 have been widely used. One problem which received attention was the nature of the product obtained when an unsymmetrical reagent such as HCl reacts with an unsymmetrical alkene such as propene. We may imagine two products from this reaction:

$$CH_3{-}CH{=}CH_2 + HCl \longrightarrow CH_3{-}CH_2{-}CH_2Cl \quad \text{(1-chloropropane)}$$

$$\text{or } CH_3{-}CH{=}CH_2 + HCl \longrightarrow CH_3{-}\underset{\displaystyle Cl}{\underset{|}{C}H}{-}CH_3 \quad \text{(2-chloropropane)}$$

These two alternatives are not difficult to distinguish; for example, treatment with NaOH followed by oxidation yields a carboxylic acid in the first case, and a ketone in the second; these differ in their reactions with $NaHCO_3$ and 2,4-dinitrophenylhydrazine. As a result of an extensive series of investigations, the generalisation known as Markovnikov's rule was formulated. According to this rule the hydrogen atom becomes attached to the carbon atom with the larger number of hydrogen atoms bonded to it, and the halogen atom to the carbon with the smaller number of hydrogens. Care is required in counting hydrogen atoms; in propene the carbon atom labelled α has two hydrogens attached to it (see figure 3.1), whereas that labelled β has only one (not four). Thus the reaction of propene with hydrogen chloride gives 2-chloropropane as the major product.

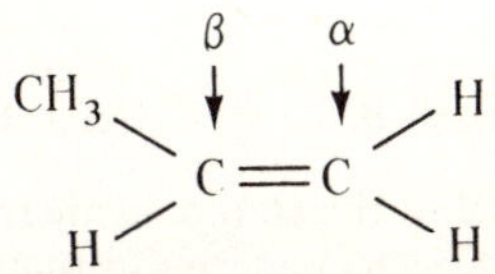

3.1 Propene

Considerable difficulties were encountered in the experimental investigations of the application of Markovnikov's rule to hydrogen bromide. Not only were different experimenters unable to agree what product was formed when propene reacted with HBr, but individual experimenters found that their results seemed to vary from day to day! It was eventually discovered that if the reagents were freshly distilled, then 2-bromopropane was formed in agreement with Markovnikov's rule, but that if they were allowed to stand in air for a few days, then the 'anti-Markovnikov' product 1-bromopropane was more likely. Dissolved oxygen was shown to be the cause of this difference.

As the addition of trace amounts of peroxides also produces 1-bromopropane, it is sometimes known as the peroxide effect. Mention of the influence of peroxides immediately makes us suspect that anti-Markovnikov addition is the result of a free radical reaction; it is therefore no surprise to us to learn that anti-Markovnikov addition also occurs in the presence of ultra-violet light.

Markovnikov addition. If we are to understand why Markovnikov's rule predicts correctly the product of the reaction of propene and hydrogen bromide, we must first be clear that there is no chemical equilibrium between 1- and 2-bromopropane in which each might be converted into the other. If this were the case the more stable isomer would always be formed, and the reaction would be said to be thermodynamically controlled. Instead the reaction is kinetically controlled, and the major product is simply the isomer which is formed more rapidly, and therefore in greater quantity. Now the rates of formation of 1- and 2-bromopropane are determined by the activation energies of the reactions:

$$CH_3{-}CH{=}CH_2 + HBr \longrightarrow CH_3CH_2CH_2Br$$

and $$CH_3{-}CH{=}CH_2 + HBr \longrightarrow CH_3CHBrCH_3$$

The reaction with the lower activation energy will have the higher rate.

The mechanism of addition reactions to alkenes have been studied in detail, and so it is now possible to understand the factors which affect the activation energies of these reactions. We need not rehearse here all the evidence in favour of the mechanism - these details may be found in the books suggested in the Further Reading. The first step is the addition of a positive ion to the alkene; in the case of HBr an H^+ ion is added:

eg $$CH_2{=}CH_2 + H^+ \longrightarrow CH_3{-}CH_2^+$$

This is then followed by attack by a negative ion (Br^- in the case of HBr) forming the addition product:

$$CH_3{-}CH_2^+ + Br^- \longrightarrow CH_3{-}CH_2Br$$

Various details are also known of the stereochemistry of these steps, but these need not concern us here. The position of highest energy is very close to the point where the H^+ ion has been added, and so the activation energy is the difference in energy between the original state, $CH_2{=}CH_2 + HBr$, and the transition state, $CH_3{-}CH_2^+ + Br^-$. This is shown in figure 3.2.

Now if we consider an addition reaction to propene, there are clearly two possible initial steps:

$$CH_3{-}CH{=}CH_2 + H^+ \longrightarrow CH_3{-}CH^+{-}CH_3$$

and $$CH_3{-}CH{=}CH_2 + H^+ \longrightarrow CH_3{-}CH_2{-}CH_2^+$$

Attack by Br^- will then yield $CH_3CHBrCH_3$ and $CH_3CH_2CH_2Br$ respectively, and this is shown in figure 3.3. The problem of comparing activation energies

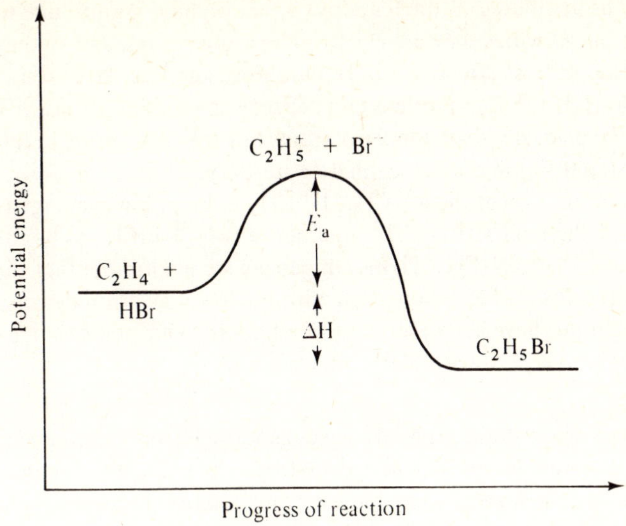

3.2 The addition of HBr to ethene

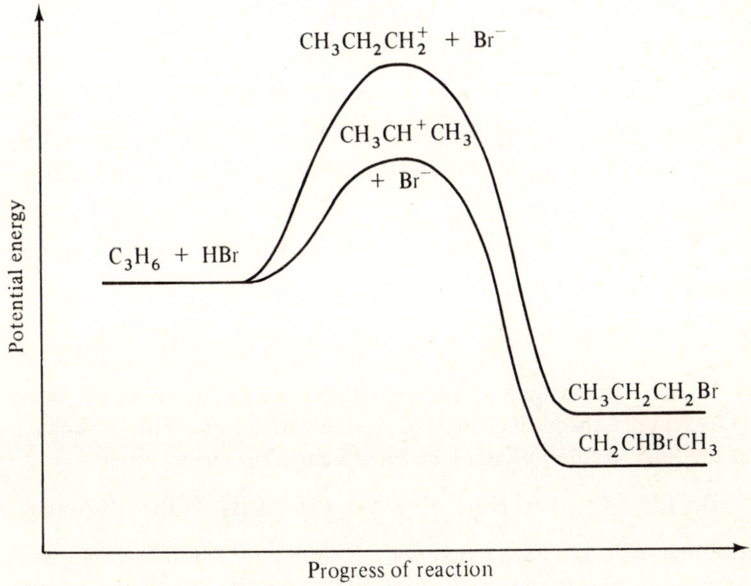

3.3 The addition of HBr to propene

for Markovnikov and anti-Markovnikov addition is now reduced to that of comparing the stabilities of the ions $CH_3CH^+CH_3$ and $CH_3CH_2CH_2^+$. The stabilities of such ions are in the order primary < secondary < tertiary, where primary ions have the formula RCH_2^+, secondary RCH^+R, and tertiary R_3C^+.

This may be attributed to the inductive effect of alkyl groups; due to their low electronegativities they are electron-donating groups, and so they spread positive charge most effectively in tertiary ions, and least effectively in primary ions. If the R groups contain benzene rings, then the positive charge may also be delocalised via the π electrons, but this still results in the same order of stability primary < secondary < tertiary.

We therefore expect the ion $CH_3CH^+CH_3$ to be of lower energy (more stable) than the ion $CH_3CH_2CH_2^+$, and hence $CH_3CHBrCH_3$ to be formed more rapidly than $CH_3CH_2CH_2Br$. (In passing we might note that the H^+ ion actually adds at the carbon atom with the lower electron density, which might not have been everyone's first guess!) Our prediction is now satisfactorily in accord with Markovnikov's rule.

Anti-Markovnikov addition. As we have seen, the introduction of small amounts of peroxides into an alkene–hydrogen bromide mixture produces a change in mechanism and a free radical chain reaction becomes dominant. Notice that the ionic mechanism is not prevented in any way: it is simply that it is much slower than the radical mechanism, and therefore has a negligible influence on the nature of the product.

The initial step is the formation of a reactive radical. This may come from the decomposition of a peroxide:

eg $$R{-}O{-}O{-}R \longrightarrow R{-}O\cdot + \cdot O{-}R$$

It may also be produced by the action of ultra-violet light. This radical then abstracts a hydrogen atom from an HBr molecule, and the chain reaction begins:

$$R{-}O\cdot + HBr \longrightarrow R{-}OH + Br\cdot \qquad \text{(Initiation)}$$

The bromine atom reacts with an alkene molecule giving:

$$Br\cdot + CH_2{=}CH_2 \longrightarrow Br{-}CH_2{-}CH_2\cdot \qquad \text{(Propagation)}$$

(The Br· may also abstract a hydrogen atom from an HBr molecule, but this produces no net chemical change.) Collision with another HBr molecule now gives the final addition product, and regenerates the chain carrier:

$$Br{-}CH_2{-}CH_2\cdot + HBr \longrightarrow Br{-}CH_2{-}CH_3 + Br\cdot \qquad \text{(Propagation)}$$

Notice that hydrogen abstraction is a very common radical reaction: we have also encountered it in the chlorination and combustion of alkanes, in polymerisation and in inhibition. The possible termination reactions are:

$$\begin{aligned} Br\cdot + Br\cdot &\longrightarrow Br_2 \\ Br\cdot + \cdot CH_2CH_2Br &\longrightarrow BrCH_2CH_2Br \qquad \text{(Termination)} \\ BrCH_2CH_2\cdot + \cdot CH_2CH_2Br &\longrightarrow Br(CH_2)_4Br \end{aligned}$$

Traces of these compounds are found in the final reaction mixture.

If we consider the radical addition of HBr to propene, then the possible first steps are:

$$CH_3{-}CH{=}CH_2 + Br\cdot \longrightarrow CH_3{-}CH\cdot{-}CH_2Br$$
$$CH_3{-}CH{=}CH_2 + Br\cdot \longrightarrow CH_3{-}CHBr{-}CH_2\cdot$$

The shapes of the potential energy curves are not quite the same as in figure 3.3, but the same general principles apply, and the rates of these two reactions will reflect the relative stabilities of the radicals CH_3-CH•-CH_2Br and CH_3-CHBr-CH_2•. The order of stabilities of free radicals is the same as that for the positive ions, primary < secondary < tertiary; different workers have explained this order in different ways! Nevertheless the radical CH_3-CH•-CH_2Br is the more stable, and is formed in preference to CH_3-CHBr-CH_2•. Hydrogen abstraction now regenerates the chain carrier to form the anti-Markovnikov product:

$$CH_3{-}CH\cdot{-}CH_2Br + HBr \longrightarrow CH_3{-}CH_2{-}CH_2Br + Br\cdot$$

The key difference between the free radical and ionic mechanisms is that in the former initial attack is by Br•, whereas in the latter hydrogen attacks first, as H^+.

Addition of HCl and HI. Although the addition of hydrogen bromide to alkenes proved troublesome to early researchers, surprisingly the addition of hydrogen chloride and iodide did not present similar problems; it was found that Markovnikov's rule was always obeyed, even under conditions which favour radical reactions. The reason for this is that with HBr radical addition is a very fast reaction, and swamps the ordinary ionic addition; for HCl and HI the reverse is true, and ionic addition is always the faster reaction.

As we have seen, the radical addition of hydrogen bromide to alkenes proceeds by a chain reaction, in which the propagation steps are:

$$CH_2{=}CH_2 + Br\cdot \longrightarrow \cdot CH_2{-}CH_2Br$$
$$\cdot CH_2{-}CH_2Br + HBr \longrightarrow CH_3{-}CH_2Br + Br\cdot$$

These are consecutive steps, and the rate of the chain reaction is therefore controlled by whichever of these reactions is the slower. For the chain reaction to be rapid, it is necessary that both propagation steps should have high reaction rates, and hence both must have low activation energies. We may get some idea of the activation energies for these steps by calculating their enthalpy changes; remember that the activation energy (E_a) must always be more positive than ΔH. Table 3.3 shows the enthalpy changes for the propagation steps for the addition of HCl, HBr and HI to an alkene.

Table 3.3 *Enthalpy changes during the addition of HCl, HBr and HI to an alkene (ΔH measured in kJ mol^{-1})*

	HCl	*HBr*	*HI*
$RCHCH_2 + X\cdot \longrightarrow R\dot{C}HCH_2X$	−76	−19	+40
$R\dot{C}HCH_2X + HX \longrightarrow RCH_2CH_2X + X\cdot$	+42	−23	−91

The high bond energy of HCl makes the second step very unfavourable; its activation energy must be greater than 42 kJ mol^{-1}, and the step is therefore very slow. In contrast, for HI the weakness of the C–I bond means that the reaction $RCHCH_2 + I \longrightarrow RCHCH_2I$ has a high activation energy, and this step now prevents the chain reaction from occurring. Only for HBr are the bond energies such that both propagation steps have low activation energies, and only in HBr is radical addition much faster than the ionic reaction.

Radical addition to alkenes is not restricted to HBr: H_2S also adds in the same way, and the reaction of propene with H_2S gives $CH_3CH_2CH_2SH$ under suitable conditions. The halogens chlorine and bromine can also add by a radical mechanism, although here Markovnikov's rule considerations no longer apply.

D Addition to benzene

It is well known that the main reaction characteristic of benzene compounds is substitution, and not addition as in the alkenes. Thus chlorine reacts with benzene in the presence of a catalyst such as $AlCl_3$, forming chlorobenzene:

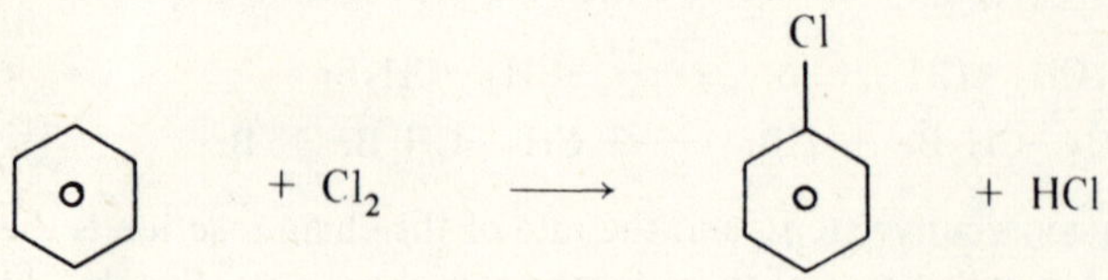

Just as with the alkenes, the initial attack is by a positive ion; in this case an incipient Cl^+ ion is formed by the interaction of Cl_2 with the catalyst:

$$AlCl_3 + Cl_2 \longrightarrow AlCl_4^- + Cl^+$$

Whether the $AlCl_3$ actually splits the chlorine forming independent ions, or merely polarises the molecule is not important here. The Cl^+ then reacts with a benzene molecule:

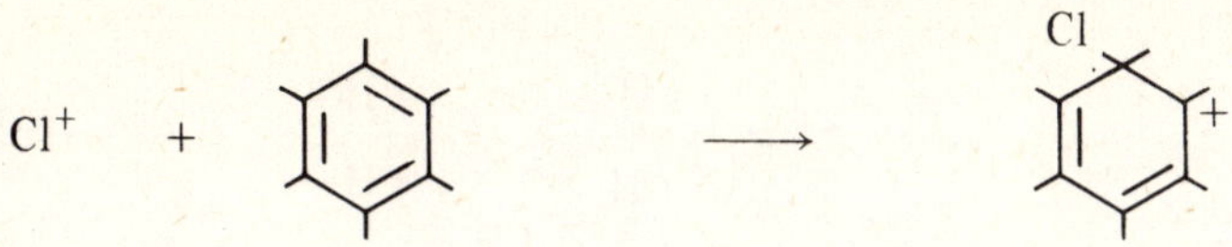

Although this could now pick up a Cl^- ion forming an addition product, it is energetically more favourable to lose an H^+ ion, restoring the stability of the benzene ring:

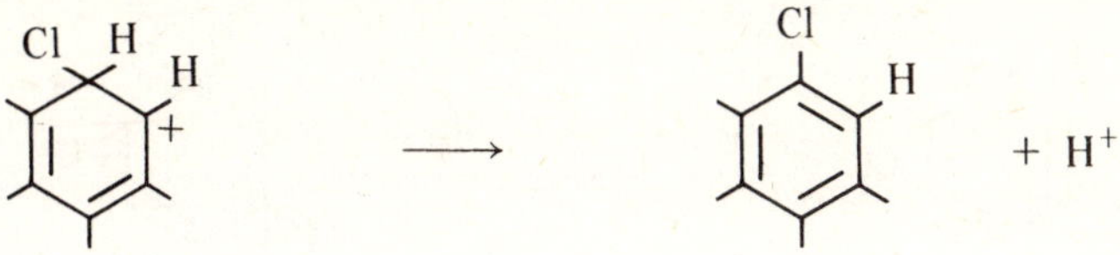

However if a mixture of benzene and chlorine is irradiated with strong light, the products are entirely different; a chain reaction occurs, giving rise to addition. The initiation step is familiar from the chlorination of alkanes:

$$Cl_2 + \text{light} \longrightarrow Cl\cdot + Cl\cdot \qquad \text{(Initiation)}$$

These chlorine atoms add across a benzene double bond, giving:

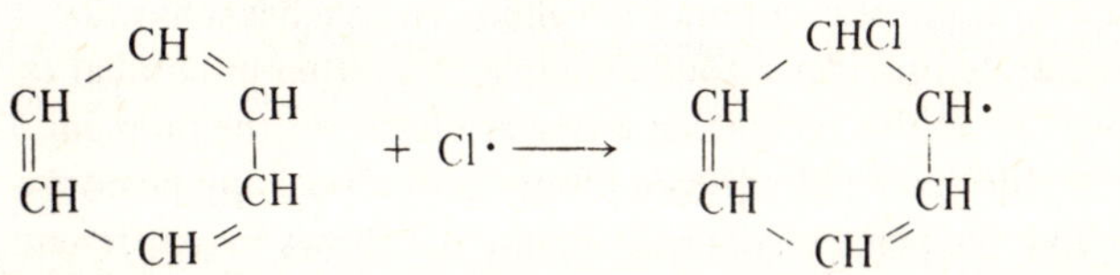

(Propagation)

The alternative reaction in which a hydrogen is abstracted is too endothermic to be competitive:

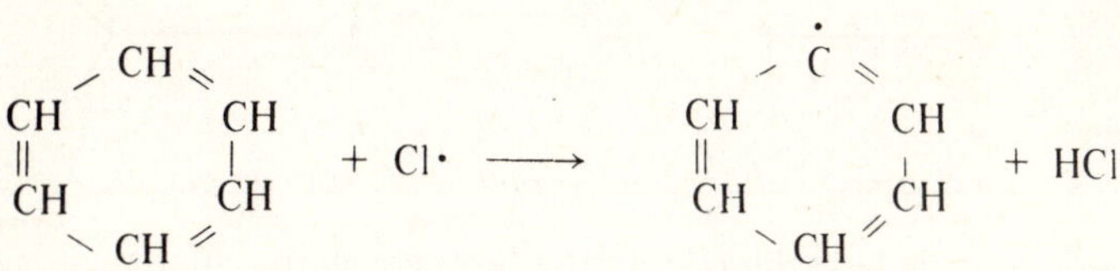

This reaction would have produced the substitution product chlorobenzene. The chain reaction then continues with the addition of another Cl atom which regenerates the chain carrier, $Cl\cdot$:

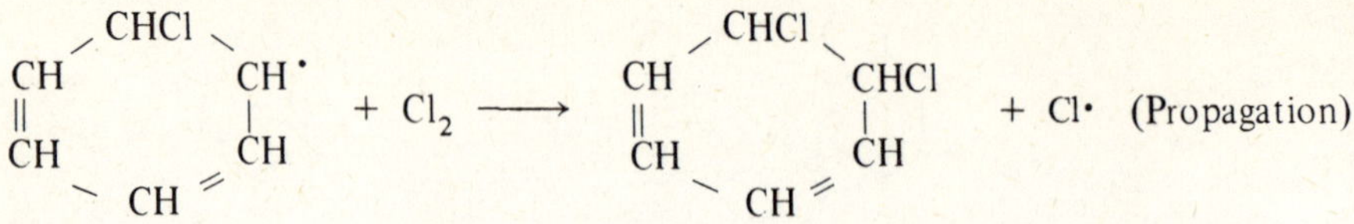

In this example termination reactions are of no great interest, and the prediction of the trace products formed by termination may be left to the reader. The product may now undergo further addition across the two remaining double bonds, and so eventually $C_6H_6Cl_6$ is formed:

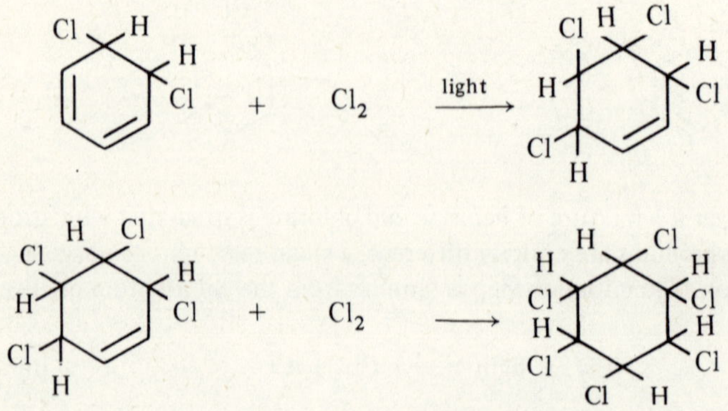

In practice it is not possible to stop the reaction after the addition of just two or four chlorine atoms. The systematic name for the product is 1,2,3,4,5,6-hexachlorocyclohexane, but it is commonly called benzene hexachloride.

Benzene hexachloride provides a good example of a rather unfamiliar type of isomerism. If we regard the cyclohexane ring as planar for the moment, then we can see that the chlorine and hydrogen atoms must all occupy positions either above or below the plane of the ring. Figure 3.4 shows two different

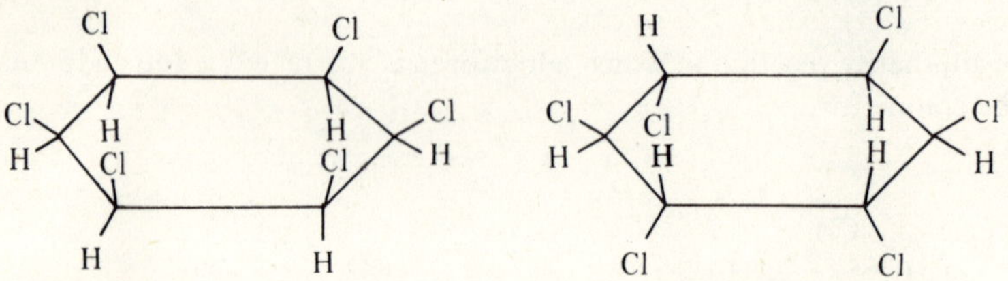

3.4 Two isomers of benzene hexachloride (shown planar)

isomers of benzene hexachloride. In the first case all the chlorine atoms are above the plane of the ring, whereas in the second three chlorines are above the plane, and three below. Altogether there are eight such isomers. The most important is γ-benzene hexachloride, a powerful insecticide sold as 'gammexane' it is formed in a yield of 10–20% when benzene and chlorine react. Of course there are other isomers of benzene hexachloride such as 1, 1, 2, 2, 3, 3-hexachlorocyclohexane, but these are not formed by the addition of chlorine to benzene, as they would require a hydrogen atom to migrate from one carbon atom to another.

In point of fact the cyclohexane ring is not planar, as this would require each C–C–C bond angle to be 120°, instead of the tetrahedral bond angle of 109°. However this makes no difference to the total number of isomers. Figure 3.5 shows the correct conformation of the cyclohexane ring; the particular isomer shown here is γ-benzene hexachloride.

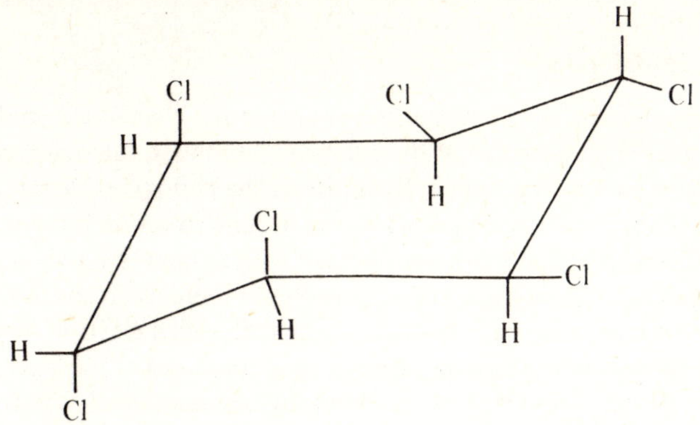

3.5 γ-benzene hexachloride

4 Ionic compounds: the main group elements

A Introduction

Ionic compounds are the usual starting point for any elementary study of valence theory. The formulae of the chlorides, bromides, sulphides and nitrides of the metals of groups I, II and III of the periodic table reveal obvious patterns; with a little knowledge of atomic structure it is not difficult to conclude that when metals react with non-metals, they do so in proportions which allow the metal to give one or more electrons to the non-metal, so that ultimately each has only completely-filled electron shells. The resulting ions then have the same electron configurations as the inert gases. Thus the reaction of lithium with sulphur may be represented as in figure 4.1.

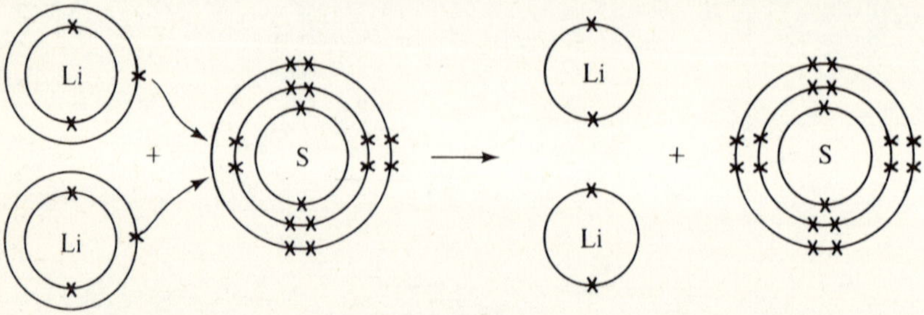

4.1 $2Li + S \longrightarrow 2Li^+ + S^{2-}$

As a theory of valency, this has two major disadvantages. Firstly it relies heavily on a suitable choice of example; thus although the series NaCl, $MgCl_2$, $AlCl_3$ suggests a simple pattern in formulae, the same could not be said of the series Li_2O, Na_2O_2, KO_2, nor of the transition metal compounds $FeCl_2$, $FeCl_3$, $CoCl_2$, CuCl, $CuCl_2$. Secondly the theory provides no explanation of why the electron configurations of the inert gases should be so common; as a result, they are often thought of as having some almost mystical significance. This gives rise to some remarkably unhelpful explanations: for example it is sometimes said that 'a sodium atom wants to get rid of an electron'. Even if we ignore the objection that desire is not a property of atoms, the statement scarcely tallies with the experimental fact that the removal of an electron from a sodium atom actually requires some 500 kJ mol^{-1} of energy.

In general terms, the main group elements (ie those which are not transition metals) tend to obey the simple rules of valency, whereas the transition metals do not. Accordingly we shall treat transition metals entirely separately in chapter 5; in this chapter we shall consider the ionic compounds of the main group elements in which unpaired electrons are found; these will of course be exceptions to the simple valency rules. In this way we shall hope to understand not only these exceptions, but also why it is that inert gas configurations are so important in the chemistry of the main group elements.

B Born–Haber cycles

Any discussion of the stability of an ionic compound must focus on its heat of formation, that is the enthalpy change when it is formed from its constituent elements, each in their standard (most stable) state. In general we may expect that if the heat of formation is negative the compound will not spontaneously decompose into its elements, whereas if the heat of formation is positive decomposition is likely, and it is therefore improbable that the compound will be able to be isolated. (Strictly speaking we should be evaluating ΔG_f, the free energy of formation. In the majority of ionic compounds the difference between ΔG_f and ΔH_f is relatively small, and is not normally chemically decisive.)

An understanding of the factors which influence the sign of the heat of formation of an ionic compound may be obtained from a Born–Haber cycle. Here the reaction of a metal and a non-metal is broken down into a series of simple steps, where the enthalpy change of each step is known. By the law of conservation of energy, often referred to as Hess' law when used in chemical calculations, the heat of formation is equal to the sum of the enthalpy changes of all the constituent steps. Figure 4.2 shows the Born-Haber cycle for NaCl. Starting from solid sodium and chlorine gas, the first two steps are to convert these elements into gaseous atoms. For sodium the enthalpy change is the latent heat of sublimation (or atomisation energy, $E_{at(Na)}$), and for chlorine it is half the Cl–Cl bond energy (half because breaking one bond produces two Cl atoms ($E_{at(Cl)}$)). The sodium atoms then each lose one electron becoming Na^+(g), and the energy required for this is the first ionisation energy of Na ($E_{i(Na)}$). The electrons then combine with the chlorine atoms, evolving an amount of energy called the electron affinity (A). The final step is the bringing together of the gas-phase Na^+ and Cl^- ions to form solid NaCl; the enthalpy change for this step is the lattice energy (E_l). Of course it would not be practically possible to convert sodium and chlorine into sodium chloride in this way, but that is not important; we need only to be able to measure the enthalpy changes for each of these steps separately. Their sum must then equal the heat of formation of NaCl.

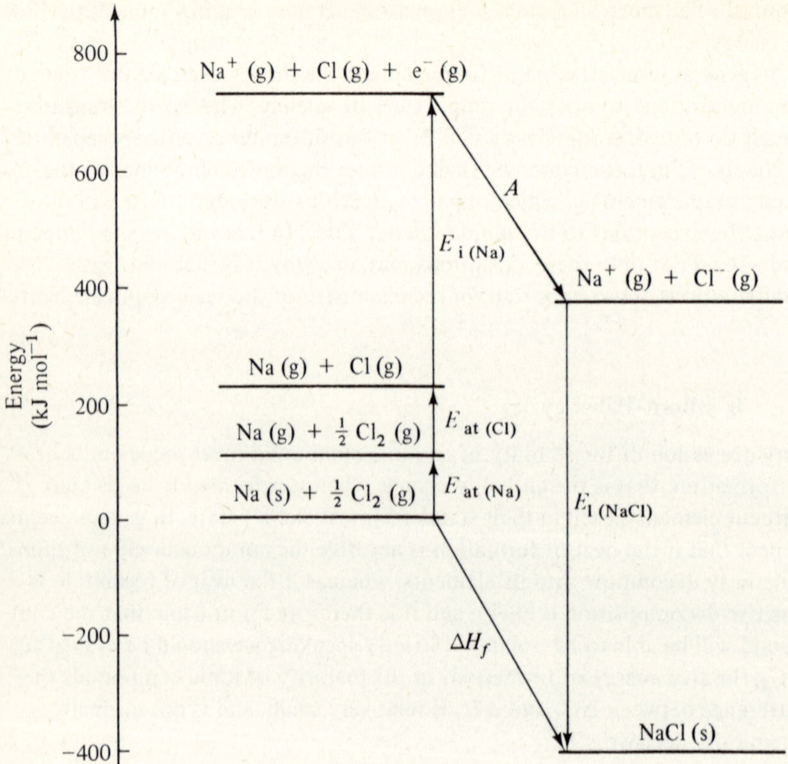

4.2 *Born–Haber cycle for NaCl*

Table 4.1 *Born–Haber cycle for NaCl*

$Na(s)$	$\longrightarrow$	$Na(g)$	$+108\ kJ\ mol^{-1}$
$\frac{1}{2}Cl_2(g)$	$\longrightarrow$	$Cl(g)$	$+121$
$Na(g)$	$\longrightarrow$	$Na^+(g) + e^-$	$+500$
$Cl(g) + e^-$	$\longrightarrow$	$Cl^-(g)$	-364
$Na^+(g) + Cl^-(g)$	$\longrightarrow$	$NaCl(s)$	-776
$Na(s) + \frac{1}{2}Cl_2(g)$	$\longrightarrow$	$NaCl(s)$	-411

Table 4.1 gives the values of the enthalpy changes for each step in the Born-Haber cycle for sodium chloride. All the values may be determined experimentally except for the lattice energy; here the value of the lattice energy has been found by using the law of conservation of energy. However calculation of the lattice energy is possible if the radii of the ions Na^+ and Cl^- are known; it simply requires a summation of all the interaction energies between

the different ions in the crystal. We assume that the interaction between ions is electrostatic, and so the interaction energy of any two ions is $(q_1q_2/4\pi\epsilon_0 d)$, where the ionic charges are q_1 and q_2, and the distance between them is d. Summing all these energies is a tedious but straightforward business, and it is found that for NaCl the calculated lattice energy is in good agreement with that obtained from the Born–Haber cycle. This agreement we may take as confirming our view that in sodium chloride the lattice is indeed held together by electrostatic forces.

In principle the determination of any five of the energy changes involved in a Born–Haber cycle allows the sixth to be calculated. In practice there are three main ways in which Born–Haber cycles are used. Firstly they may be used to compare experimental and calculated lattice energies, as for NaCl above, in order to investigate whether the binding of the ions may be quantitatively described as the electrostatic interactions of charged particles. This sort of comparison has been used to show that there is appreciable covalency in solids such as ZnS. Secondly if we assume that a calculated lattice energy is correct, we may deduce the value of the electron affinity of the non-metal in an ionic compound. This has been of value in some cases, as some electron affinities are very difficult to determine experimentally. Lastly Born–Haber cycles may be used to predict the values of the heats of formation of unknown compounds; these values may then be used to explain why these compounds are unknown. In the next section we shall use Born–Haber cycles to explain why it is that MgF and MgF_3, both odd-electron compounds, have never been prepared in spite of extensive efforts.

There are two features of the Born–Haber cycle for sodium chloride which we should note before we pass on to other examples. Firstly the transfer of an electron from a sodium atom to a chlorine atom is an endothermic process, ie in energy terms is unfavourable. This is not only true for NaCl but for all metal-non-metal pairs; this casts a revealing light on the idea that atoms lose and gain electrons in order to acquire an inert gas configuration. Clearly the stability of NaCl has more to do with the size of its lattice energy than with any supposed desirability of closed electron shells. Secondly we should note that the two largest terms in the Born–Haber cycle are the ionisation energy of Na and the lattice energy of NaCl; the other terms are relatively small. This is a feature of most Born–Haber cycles, and the sign of ΔH_f for an ionic compound tends to be determined by the relative magnitudes of the ionisation and lattice energies.

C MgF and MgF_3

We may now proceed to calculate the heats of formation of the unknown compounds MgF and MgF_3. In the first of these the charge on the Mg ion is $+1$, corresponding to the electron configuration $1s^2 2s^2 2p^6 3s^1$, with one electron in the outer shell; in the second the Mg ion has the charge $+3$, with electron configuration $1s^2 2s^2 2p^5$ and only seven electrons in the outer shell.

For comparison purposes we shall also need to calculate the heat of formation of MgF_2, which is invariably formed when magnesium burns in fluorine.

Table 4.2 compares the Born–Haber cycles for MgF, MgF_2 and MgF_3. Most of the terms are straightforward and require no comment. However a little care is needed in multiplying the electron affinity and atomisation energy of fluorine by two and three respectively in MgF_2 and MgF_3; similarly the energy required to remove two electrons from a Mg atom is the sum of the first and second ionisation energies. Choosing values for the lattice energies of MgF and MgF_3 is more difficult: they are not known experimentally, as neither compound has been synthesised, nor are the exact radii of the Mg^+ or Mg^{3+} ions known. But we can make some reasonable guesses based on related compounds, and as it turns out the precise values will not be very important. The lattice energy of a crystal is inversely proportional to the interionic distance, if we compare solids with the same ionic charges and crystal structure. Now an Mg^+ ion must have a larger radius than an Na^+ ion, as it has an extra shell occupied, and therefore we may expect the lattice energy of MgF to be less negative than that of NaF, which is known to be −915 kJ mol^{-1}. In contrast an Mg^{3+} ion is likely to be larger than an Al^{3+} ion, as the same shells are occupied, but the nuclear charge is greater in Al, and so the lattice energy of MgF_3 must be less negative than that of AlF_3, which is known to be −5950 kJ mol^{-1}. The Born–Haber cycles are shown diagrammatically in figure 4.3. They confirm that the values of the heats of formation are in each case determined mainly by the ionisation energies of the Mg atom, and the lattice energy of the solid: other terms are relatively small.

Table 4.2 *Born–Haber cycles for MgF, MgF_2 and MgF_3*

Enthalpy changes (kJ mol^{-1})	*MgF*	*MgF_2*	*MgF_3*
Mg atomisation	+150	+150	+150
F atomisation	+80	+160	+240
Mg ionisation	+740	+2240	+9940
F ionisation	−330	−670	−1000
Lattice energy	−900 (?)	−2880	−5900 (?)
Heat of formation	−260	−1000	+3430

It is immediately obvious from figure 4.3 why MgF_3 has never been prepared: its heat of formation is large and positive, and we should expect it to decompose to its elements readily. Notice that we have used a value of the lattice energy which is certainly too negative; in reality the value of ΔH_f would be even more positive. The reason why ΔH_f is so unfavourable is that the third ionisation energy of Mg (the energy required to convert Mg^{2+} to Mg^{3+}) is extremely high, 7700 kJ mol^{-1}. Although the lattice energy of MgF_3 is greater than that of MgF_2, due to the higher ionic charge, this is

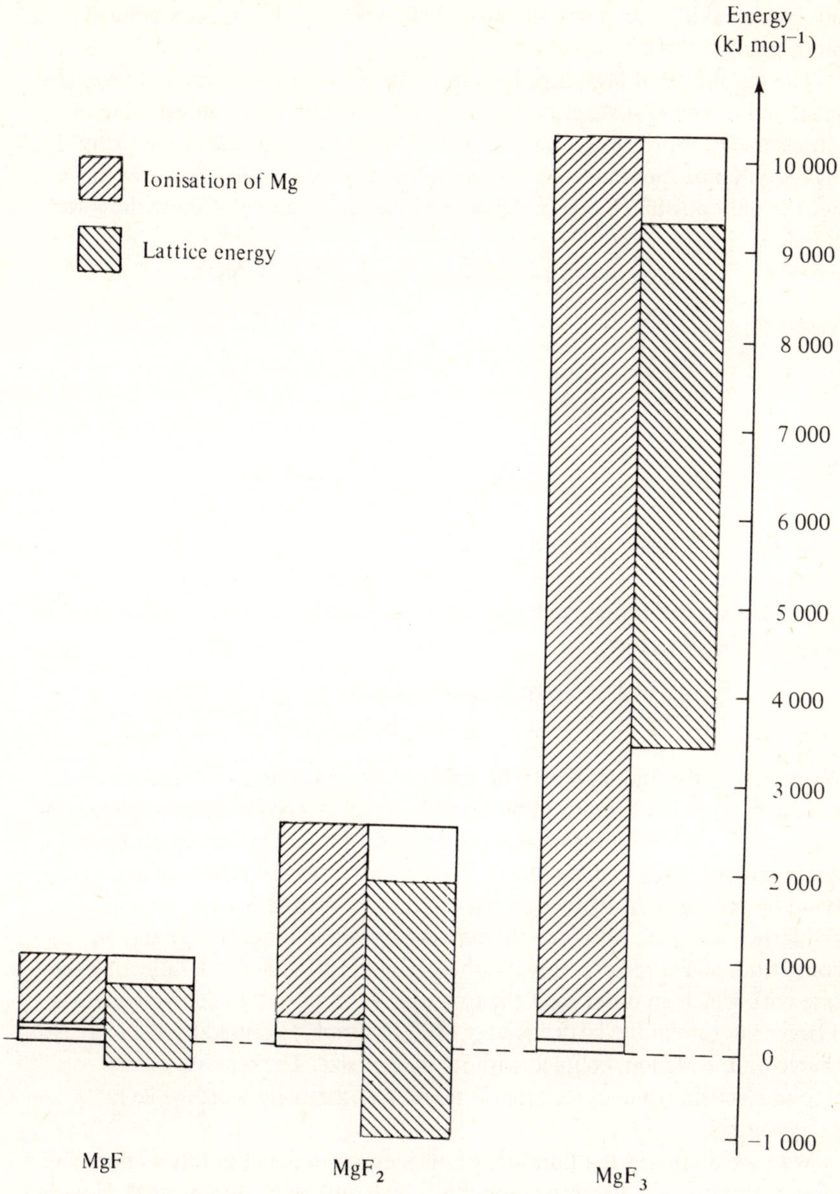

4.3 Born–Haber cycles for MgF, MgF_2 and MgF_3

quite insufficient to compensate for the enormous increase in ionisation energy. The high ionisation energy is in turn attributable to the fact that it involves breaking into a new electron shell, where the electrons are much more tightly bound.

The instability of MgF is rather more subtle. Its heat of formation we have predicted to be negative, although we must remember that our estimate of the lattice energy is probably over-charitable. Nevertheless it does seem likely that MgF would not spontaneously decompose into its elements. However this is not the only possible mode of decomposition, and figure 4.4 shows how we

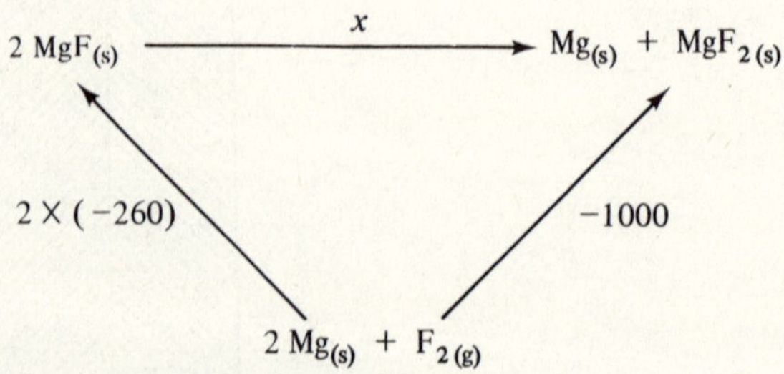

4.4 The decomposition of MgF

may calculate the enthalpy change for a more likely process. By the law of conservation of energy:

$$(-2 \times 260) + x = -1000$$

Therefore $$x = -480\ \text{kJ mol}^{-1}.$$

We therefore see that MgF may be stable to decomposition into its elements, but decomposition forming MgF_2 and Mg is still a very favourable reaction. If magnesium and fluorine are mixed in the ratio 1 : 1, it is found that the product contains MgF_2 and Mg metal. The extra stability of MgF_2 we may understand by reference to figure 4.3; once again the main terms are the ionisation and lattice energies. Although there is an increase in ionisation energy in going from MgF to MgF_2, this is rather small (1500 kJ mol^{-1}), reflecting the ease with which an outer shell electron may be removed. In contrast there is a larger increase in the lattice energy (2000 kJ mol^{-1}) caused by the increased charge on the Mg ion, coupled with its smaller size. The removal of the second electron from an Mg atom is therefore extremely worthwhile in energy terms.

We have discussed the fluorides of magnesium in detail in this section, and shown why the odd-electron compounds MgF and MgF_3 do not exist. However we could have chosen almost any metal from groups I–III, and any non-metal from groups V–VII, and formed the same conclusions. The outer shell electrons of the metals have relatively low ionisation energies, and it is always worthwhile to remove all of them, as this gives high lattice energies. In contrast the electrons from the inner shells are much more difficult to remove, and the extra lattice energy obtained from the higher ionic charge is never sufficient compensation for the enormous ionisation energies. Figure 4.5 shows the variation in successive ionisation energies across the periodic table; in each case there is a marked jump after the outer shell electrons have been removed.

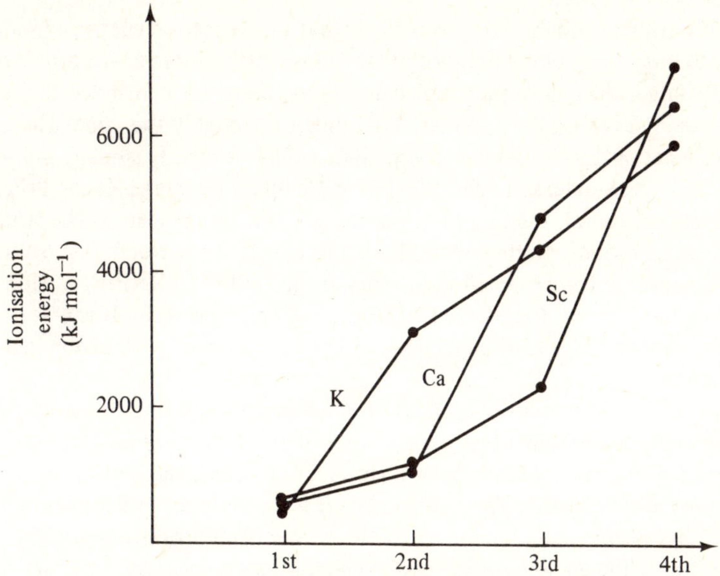

4.5 Successive ionisation energies of K, Ca and Sc

So far we have been concerned with explaining why stable odd-electron species should not be formed when the main group elements react with each other. For the rest of this chapter we shall look at some of the more important exceptions to this rule.

D The discovery of inert gas compounds

One of the most dramatic applications of a Born–Haber cycle was the discovery of the first inert gas compound by Bartlett in 1962. Until then it was believed that the inert gases formed no true compounds; only clathrates, in which the inert gas atoms were trapped in spaces between molecules in a solid lattice, were known, but these do not involve any chemical binding.

Bartlett was investigating the properties of a new oxidising agent of remarkable strength, platinum hexafluoride (PtF_6). He was able to show that it could form an ionic compound by reacting with oxygen gas:

$$O_2 + PtF_6 \longrightarrow O_2^+ + PtF_6^-$$

This is in itself a quite remarkable reaction; both the ions formed are odd-electron species. The ionisation energy of an oxygen molecule ($O_2 \longrightarrow O_2^+$)

is known from molecular spectroscopy to be very high, about 1200 kJ mol^{-1}. In this case it seems that the heat of formation is negative because of the extremely high, although unknown, electron affinity of PtF_6. This is in contrast to the other ionic compounds which we have studied, where the electron affinity was generally a small term compared with the ionisation and lattice energies.

With brilliant insight, Bartlett realised that the ionisation energy of xenon, 1170 kJ mol^{-1}, is comparable with that of O_2; furthermore a Xe atom is only slightly larger than an oxygen molecule, and he therefore concluded that the lattice energies of $O_2^+PtF_6^-$ and $Xe^+PtF_6^-$ might be roughly the same. The Born–Haber cycles for the two compounds would be closely similar, and so if O_2PtF_6 was stable, so might $XePtF_6$ be. He therefore mixed Xe and PtF_6 and obtained a red solid, the first inert gas compound to be prepared. The structure of xenon hexafluoroplatinate is still not completely understood; one of the problems is that its composition can vary from $XePtF_6$ to $Xe(PtF_6)_2$. However this does not detract from the merit of the original suggestion. It is now known that other transition metal hexafluorides can react with xenon in a similar way.

The discovery of $XePtF_6$ sparked off a major effort to find other inert gas compounds, and within a few years a series of compounds of xenon with oxygen and fluorine became known. However no other members of the group have proved so versatile. Krypton forms a few compounds, but studies of radon have been greatly hampered by the element's radioactivity, and the higher ionisation energies of the earlier elements seem to prevent any stable compounds being formed.

The discovery of xenon compounds has brought the term *inert gas* into some disfavour; however the common use of the term *rare gas* is, if anything, even less desirable, as argon constitutes almost 1% of the Earth's atmosphere. The term *noble gas* is probably the least unsatisfactory description, although it would be wrong to suggest that this usage is universal.

E The flame test

The great majority of the chemistry with which we are familiar is concerned with the properties of substances at fairly low temperatures ie less than 1000 K. This is quite sufficient for most purposes; few reactions which occur at higher temperatures are of great synthetic value, and there are severe practical problems in handling very hot substances. However there are occasions when employing very high temperatures can be of value, and one of the best-known examples is in the use of the flame test in qualitative analysis to identify metallic elements.

We must not necessarily expect the properties of materials at high temperatures to be familiar to us: changes in temperature can have a profound effect on the stability of compounds. At very low temperatures most substances are solids; if we heat them they turn into liquids and gases, and this process may involve the breaking of weak bonds such as hydrogen bonds, or of van der Waals' forces. If we continue to increase the temperature, then weak covalent bonds can be broken – the formation of radicals from benzoyl peroxide which we saw in the last chapter is a simple example. Higher temperatures cause stronger covalent bonds to break, and large molecules decompose forming smaller molecules. Eventually these small molecules also decompose, and atoms become the commonest particles. Under the extreme conditions in hot stars, electrons can be stripped from atoms, and the result is a plasma. This trend to smaller particles is quite general, and is the result of the importance of entropy effects increasing as the temperature rises.

In the flame test, a sample of a metallic compound is held on a wire in an ordinary bunsen flame, and the subsequent colour of the flame is observed. A number of metals give rise to a characteristic colouration; the best-known example is the strong yellow flame obtained from sodium compounds. Table 4.3 gives the colours caused by some common metals. The flame test has the

Table 4.3 *Flame test colours*

Lithium	Red
Sodium	Yellow, colourless through blue glass
Potassium	Violet, purple through blue glass
Caesium	Violet
Strontium	Scarlet
Calcium	Red–yellow
Copper	Green–blue

advantages of being quick, cheap, and requiring a very small amount of the substance under test. It is particularly valuable for the group I metals, which give rise to readily distinguishable colours, and which are hard to detect in other ways as they form so few insoluble compounds. With more advanced optical systems it is possible to extend the test to allow the detection of traces of metals, even in the presence of other strong light emitters.

The important point about the flame test is that the colouration is caused by neutral metal atoms, which are formed when the sample is placed in the flame; it does not arise from metal cations, which would be present in the solid sample at room temperature. This is another illustration of the fact that the transfer of an electron from a metal to a non-metal is energetically unfavourable; it occurs only when the final product is a solid, when the lattice energy makes the overall process of electron transfer followed by lattice formation worthwhile. Thus in discussing the colours of flames, we shall be concerned with free atoms and not closed shell ions.

Emission spectra. In discussing how elements come to emit light of characteristic frequencies on heating, we will select sodium atoms as our example; the yellow colouration of sodium is readily observed by allowing a pan of salted water to boil over on a gas ring, or by looking at a sodium street lamp. Remember that the emission of light is caused by the very high temperatures in a flame or lamp; it does not represent sodium burning to give sodium oxide, or any other chemical reaction.

A sodium atom has the electron configuration $1s^2 2s^2 2p^6 3s^1$; this is obtained by allowing the 11 electrons to occupy the most stable orbitals possible, subject to there being no more than two electrons in any *s* orbital, or each of the three 2*p* orbitals. The 1*s*, 2*s* and 2*p* electrons are held very strongly to the nucleus, and remain unaltered in all chemically important conditions: they are called core electrons. But the 3*s* electron is less tightly held, and is of course the electron used in chemical binding: it is called a valence electron. Even under the extreme conditions of a flame, the great majority of sodium atoms have configuration $1s^2 2s^2 2p^6 3s^1$, and are said to be in their ground state. But they are continually being bombarded by highly energetic particles, and these collisions can cause the 3*s* electron to 'jump' into another higher orbital. The most stable unoccupied orbital is 3*p*, but the electron does not necessarily jump into this orbital; it could also occupy orbitals such as 4*s*, 3*d* or 5*p*. The atom is then said to be in an excited state. However the atom does not remain in this state for very long; the electron can fall back into a more stable orbital. This is accompanied by the emission of energy in the form of light. The excitation and emission are shown in figure 4.6.

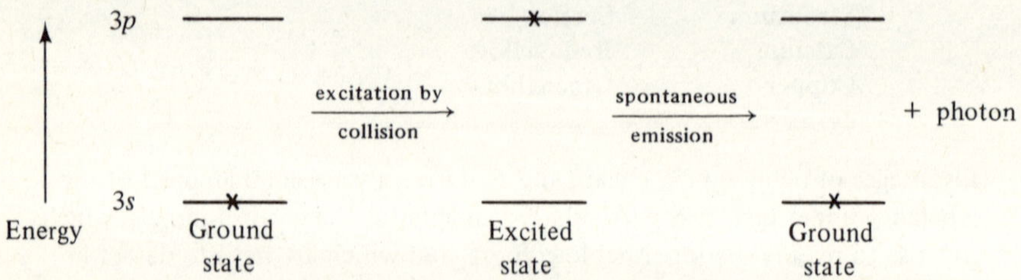

4.6 *The emission of light by a Na atom*

Once again we shall need to think of light as a stream of particles, or photons, if we are to understand why the light emitted in figure 4.6 has a particular colour and wavelength. When an atom falls from an excited state to its ground state, it decreases its energy, just as a book loses energy when it falls from a desk to the floor. In the case of the book, we know that its potential energy is converted ultimately into sound and heat energy, and that the total amount of energy in the universe remains the same. In the case of an atom, the energy lost by the electron is converted into light energy, and a photon is produced. For the particular example shown in figure 4.6, the difference in energy between the 3*p* and 3*s* orbitals is 3.36×10^{-19} J for one sodium atom, and this must equal the energy of the emitted photon. The energy (E) of a photon of light frequency ν is given by:

$$E = h \times \nu$$

where h is Planck's constant (6.63×10^{-34} J s). Therefore for the sodium atom:

$$\begin{aligned} \nu &= E/h \\ &= 3.36 \times 10^{-19}/6.63 \times 10^{-34} \quad \text{s}^{-1} \\ &= 5.1 \times 10^{14} \quad \text{s}^{-1} \end{aligned}$$

The wavelength (λ) and frequency (ν) of light are related by

$$c = \lambda \times \nu \quad \text{(where } c = \text{ speed of light)}$$

Therefore

$$\begin{aligned} \lambda &= c/\nu \\ &= 3.0 \times 10^{8}/5.1 \times 10^{14} \quad \text{m} \\ &= 5.9 \times 10^{-7} \quad \text{m} \\ &= 590 \quad \text{nm} \end{aligned}$$

Thus the light emitted has wavelength 590 nm, which lies in the yellow region of the spectrum.

We have considered so far the simplest case of excitation, in which the 3*s* electron is excited to the 3*p* orbital, and then falls back to the 3*s* orbital again, emitting yellow light; as it happens, this is much the most important process in sodium atoms. But there are other orbitals to which the electron could have been excited, and the electron need not necessarily fall back into the 3*s* orbital. (Note however that we need not be concerned with the excitation of core electrons.) Thus the initial excitation might have produced a 4*s* electron, which could then have fallen back into the 3*p* orbital; this jump gives rise to a photon of wavelength 1140 nm, which is in the infra-red region of the spectrum. Clearly there are many such possibilities, and each will correspond to photons of a particular wavelength.

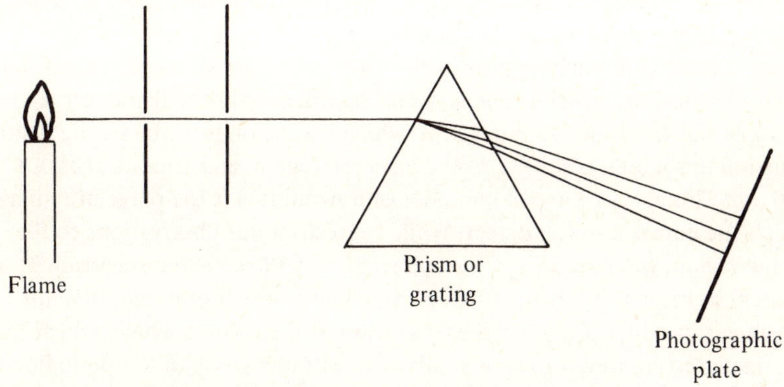

4.7 Schematic representation of a flame spectrometer

The light emitted by a sodium flame may be investigated by the apparatus shown in figure 4.7; the light is split into its component colours by a prism or diffraction grating, and may then be detected either visually or on a photo-

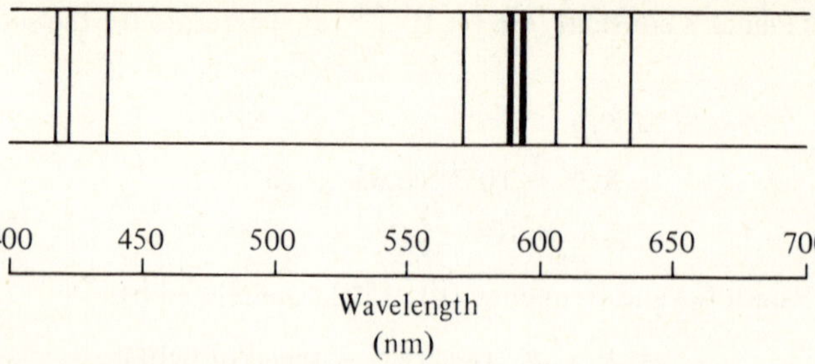

4.8 The emission spectrum of Na

graphic plate. The photograph which is obtained from a sodium flame is shown in figure 4.8: this figure is really a photographic negative, as the strong light emission at 590 nm appears as dark lines on a white background.

The sodium spectrum is quite typical of the spectra produced by most metallic elements. The appearance of individual lines in the spectrum is a characteristic of atoms: molecules and small incandescent particles give rise to spectra in which there is continuous emission of light over a range of wavelengths. The colour which the eye detects depends not only on which wavelengths of light are emitted, but also on their relative intensities; in the case of sodium, the line at 590 nm is much more intense than all the others, and so sodium light appears yellow. In a few cases the use of filters is helpful in the flame test: these cut out all light in some parts of the visible region. Thus the most convenient way of detecting potassium in the presence of sodium is to use a blue filter, which cuts out the strong yellow emission, but allows the violet light from the potassium atoms to pass through.

From figure 4.8 it would appear that there is no real pattern underlying the wavelengths at which lines appear in the sodium spectrum. Nor is there any apparent similarity between the sodium spectrum and that of potassium. This is partly because we are considering only the visible region of the spectra; many other lines appear in the ultra-violet and infra-red regions, but of course these are invisible and so of no significance in the flame test. For example the 4*p*–4*s* line in potassium, which corresponds to the strong yellow emission in sodium, is shifted to the infra-red region, and appears at about 800 nm. This acts as a useful reminder of how limited is the range of radiation which the human eye can detect. While we restrict our observations to the visible region, we must accept the appearance of the emission spectrum of an element as little more than a fingerprint, albeit a very useful one. It is not possible to produce any general explanation of the colours which appear in the flame test, nor even to spot similarities within a group in the periodic table. Applied scientists will not be too worried by this, and will be content to use the flame test as an invaluable tool in qualitative analysis. Purists may feel less satisfied, but will be glad to hear that we shall be able to make more sense of the question of colour when we discusss transition metals in the next chapter.

F Solvated electrons

One of the most remarkable properties of the alkali metals is their ability to dissolve in liquid ammonia, forming solutions which in some ways resemble liquid metals. We are accustomed to thinking in terms of solutions in water when we discuss solubility, but there are many important non-aqueous solvents. We are familiar with the use of diethyl ether and trichloromethane in organic chemistry; but these solvents are of rather limited value in inorganic chemistry, as they are incapable of dissolving ionic compounds. Liquid ammonia is an important example of a non-aqueous solvent which can dissolve ionic substances; ammonia may be liquefied simply by cooling it to below −33 °C.

Many analogies may be drawn between reactions which occur in ammonia and those in water. Both solvents are slightly self-ionising:

$$H_2O + H_2O \rightleftharpoons H_3O^+ + OH^-$$
$$NH_3 + NH_3 \rightleftharpoons NH_4^+ + NH_2^-$$

In water we define an alkali as a substance which dissolves giving OH^- ions; so in liquid ammonia we may define an alkali as a substance which dissolves giving NH_2^- ions. Thus $NaNH_2$ is an alkali in ammonia solution (equivalent in water: NaOH), and so is K_2NH (equivalent in water: K_2O). In water acidic solutions contain H_3O^+ ions, and so in ammonia acidic solutions contain NH_4^+ ions; thus NH_4Br and $(NH_4)_2SO_4$, which we might customarily think of as neutral, are considered acids in liquid ammonia. A typical neutralisation reaction would now be:

$$NH_4Cl + KNH_2 \longrightarrow KCl + 2NH_3$$
$$\text{acid} + \text{base} \longrightarrow \text{salt} + \text{ammonia}$$

This is a proton-transfer reaction, with a proton (H^+ ion) being transferred from NH_4^+ to NH_2^-, giving $2NH_3$; the equivalent reaction in water transfers a proton from H_3O^+ to OH^-, giving $2H_2O$.

All the alkali metals will dissolve in liquid ammonia; the resulting solutions are extremely good electrical conductors, and appear blue in colour. On evaporation the ammonia is boiled off, and the solid metal is reformed. It is now known that these solutions contain alkali metal cations, and electrons:

eg $$Na(s) \longrightarrow Na^+(am) + e^-(am).$$

The removal of an electron from an alkali metal atom is energetically unfavourable in the gas phase, as we have seen; but in solution both the metal cation and the electron are heavily solvated, ie surrounded by ammonia molecules which are tightly held by ion–dipole forces. Figure 4.9 shows the orientation of ammonia molecules round a sodium ion.

The properties of the alkali metal solutions are determined largely by the properties of the solvated electron, which we can think of as the simplest odd-electron species. The electron is highly mobile, and so alkali metal solutions

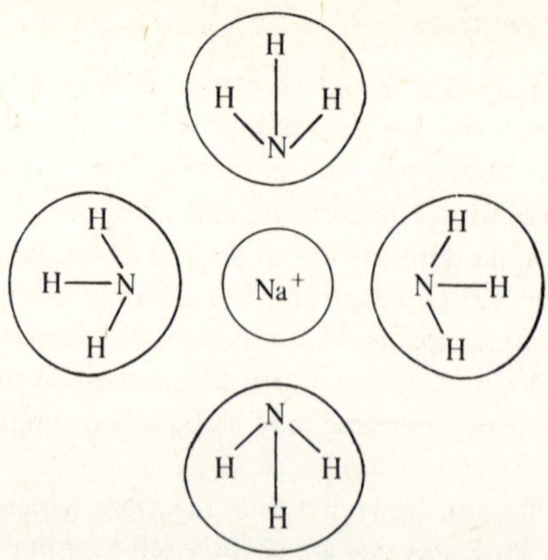

4.9 The solvation of a Na^+ *ion in liquid ammonia*

have high conductivities, rather larger than those of most alkali metal salt solutions. The solvated electron also absorbs light in the red and infra-red regions of the spectrum; it therefore transmits visible light other than red, and so appears blue to an observer. The alkali metal ions are all colourless in ammonia solution, and so the alkali metal solutions all have exactly the same blue colour.

At first sight it might seem that there are few similarities between the actions of water and ammonia on the alkali metals. Ammonia produces a solution containing electrons, whereas water reacts readily evolving hydrogen:

eg $$Na(s) + H_2O(l) \longrightarrow NaOH(aq) + \tfrac{1}{2}H_2(g)$$

The reason for this difference is the variation in stability of the solvated electron in the two solvents. It has been shown that the lifetime of an electron in water is less than one-thousandth of a second, a short time but well within the range of modern measuring techniques. It reacts as follows:

$$e^-(aq) + H_2O(l) \longrightarrow OH^-(aq) + \tfrac{1}{2}H_2(g).$$

We may therefore imagine the reaction of sodium with water as involving the formation of Na^+ ions and electrons, followed by reaction between the electrons and the solvent. In contrast, the electron is stable almost indefinitely in ammonia if the solvent is pure, and so a metal solution is formed. But if a suitable catalyst, such as a transition metal compound, is added, then decomposition is rapid:

$$e^-(am) + NH_3(l) \xrightarrow{(Fe_2O_3)} NH_2^-(am) + \tfrac{1}{2}H_2(g)$$

which is precisely analogous to the reaction in water. The overall reaction is now:

$$Na(s) + NH_3(l) \longrightarrow NaNH_2(am) + \tfrac{1}{2}H_2(g)$$

Notice that the reaction of an electron with ammonia is energetically favourable, ie the reaction is exothermic, but in the absence of a suitable catalyst the activation energy is too high for the reaction to occur at any detectable rate. The catalyst reduces the activation energy by providing an alternative reaction pathway, and decomposition then occurs rapidly.

Another way of producing hydrogen from alkali metal solutions is to add acid. We are familiar with the idea that sodium reacts much more vigorously with dilute hydrochloric acid than with water, and in exactly the same way it reacts much more vigorously with NH_4Cl in ammonia than with ammonia alone. The reaction which occurs is:

$$NH_4Cl + Na \longrightarrow NaCl + NH_3 + \tfrac{1}{2}H_2$$

The main reactions of the solvated electron are all reduction reactions: the simple addition of an electron to any ion or molecule is by definition a reduction. In fact the formation of solvated electrons in liquid ammonia (and in other non-aqueous solvents) is far from being merely a chemical curiosity. The electrons are versatile reducing agents, and are commonly used in synthetic organic chemistry, particularly with unsaturated molecules.

5 Ionic compounds: the transition metals

A Introduction

The transition metals are in many ways the forgotten elements of the periodic table. They are traditionally treated last in text-books, long after the conventional ideas of valence theory have been absorbed, when their unusual properties may be described briefly and written off as exceptions to more general rules. The quickest of glances at the periodic table reveals the inappropriateness of this view; the transition metals form the largest single group in the table, and if we include the lanthanides and actinides as transition metals, the group now includes the majority of the known elements. Furthermore the transition metals are not merely a collection of rare and valueless elements which ought properly to be left in obscurity; they include iron, nickel, copper, silver, platinum and gold, whose importance in the development of human civilisation is not easily overestimated.

The transition metals may be defined approximately as the elements from Sc to Cu, and those immediately below them in the periodic table. The characteristic properties of these elements are well known: the metals themselves have high densities, high melting points, and good mechanical and catalytic properties, and in their compounds they show variable valency, colour, complex ion formation, paramagnetism and again catalysis. In passing we note that many of their ions are odd-electron species.

All these unusual properties stem from the electron configurations of the transition metals and their ions. If we consider the first transition series from Sc to Cu, we find that the energies of the 4*s* and 3*d* orbitals are very similar. Table 5.1 shows the configurations of the neutral atoms across the series;

Table 5.1 *Electron configurations of atoms in the first transition series*

Sc	[Ar]	$4s^2 3d^1$	
Ti		$4s^2 3d^2$	
V		$4s^2 3d^3$	
Cr		$4s^1 3d^5$	
Mn		$4s^2 3d^5$	([Ar] means having the same configuration as argon)
Fe		$4s^2 3d^6$	
Co		$4s^2 3d^7$	
Ni		$4s^2 3d^8$	
Cu		$4s^1 3d^{10}$	

there is no simple pattern, and very subtle factors are responsible for the apparent anomalies at Cr and Cu. However if we consider some of the common ions of the transition metals we see a different pattern, shown in table 5.2. In the positive ions only, the 3*d* orbital is more stable than the 4*s* (detailed considerations of shape explain this unexpected difference), and so when a transition metal loses electrons, it loses its 4*s* electrons first, and any 3*d* later. As we see in table 5.2, most of the commonly occurring transition metal ions contain a *d* shell which is neither completely full nor empty, in disagreement with simple valence rules. It is these partially-filled *d* shells which give rise to many of the characteristic properties of the transition metals.

Table 5.2 *Electron configurations of ions in the first transition series*

Element	*Charges on ions*		
	1+	2+	3+
Sc			$3d^0$
Ti		$3d^2$	
V		$3d^3$	$3d^2$
Cr			$3d^3$
Mn		$3d^5$	
Fe		$3d^6$	$3d^5$
Co		$3d^7$	
Ni		$3d^8$	
Cu	$3d^{10}$	$3d^9$	

The importance of partially filled *d* shells has given rise to some discussion as to exactly which elements should be classified as transition metals. One definition restricts transition metals to those elements whose compounds contain a partially-filled *d* shell. This rules out as a transition metal scandium, whose compounds almost all contain the Sc^{3+} ion, which have the same electron configuration as Ar. It also excludes Ag, whose 1+ ion has the configuration [Kr] $4d^{10}$, in which the 4*d* shell is full. This makes good chemical sense as neither Sc nor Ag behaves as a typical transition metal. But the definition is more difficult to apply to Cu, whose ions have the configurations $3d^{10}$ (Cu^+) and $3d^9$ (Cu^{2+}). The most important point is to concentrate on the electron configurations of individual ions, rathei than to worry about whether a particular element is a transition metal or not.

B Variable valency

When we studied the main group elements in chapter 4, we came to the conclusion that the main factors influencing the heat of formation of an ionic compound, and hence its stability, are the ionisation energy of the metal and the lattice energy of the crystal; other terms are relatively small, and often roughly self-cancelling. This conclusion is generally also valid for transition metal compounds. Furthermore we concluded that the main group elements always lose all their outer shell electrons, and neither more nor less, because their outer shell electrons are easy to remove, whereas the inner shell electrons have very high ionisation energies. Thus it is always worthwhile to remove all the outer shell electrons, and never worthwhile to remove any of the inner shell electrons.

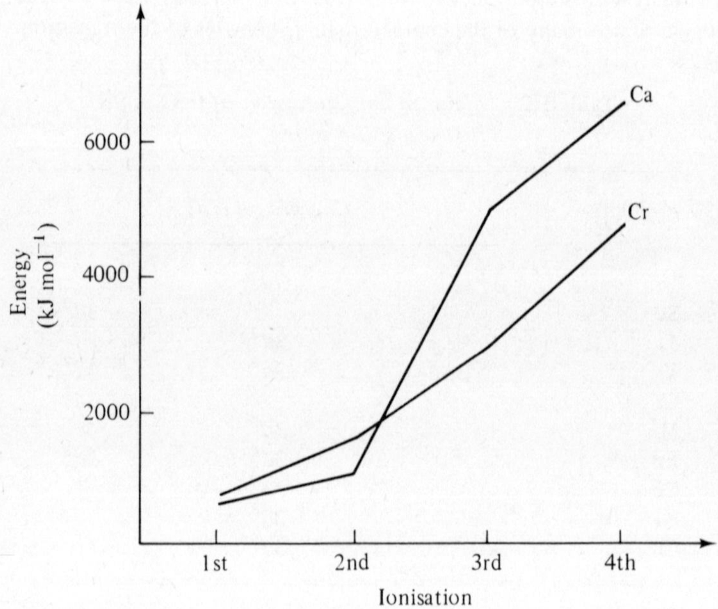

5.1 Successive ionisation energies of Ca and Cr

Figure 5.1 shows the successive ionisation energies of chromium and calcium. Two differences between the main group element, Ca, and the transition metal, Cr, are apparent. Firstly the first two ionisation energies of Cr are a little higher than those of Ca. Secondly whereas there is a definite jump to the third and subsequent ionisation energies of Ca, there is no such jump in Cr, and the ionisation energies increase steadily until all the 4*s* and 3*d* electrons (six in Cr) have been removed. Therefore there is no sudden change in stability if we consider the series of oxides CrO, Cr_2O_3 and CrO_3, in which the charge on the chromium increases, and under varying chemical conditions all these compounds may be formed.

Oxidation numbers. Before we discuss the general patterns of valency which are found among the transition metals, we must first think rather carefully about the nature of the bonding in some transition metal compounds. It is usual to talk of two types of chemical bonding, ionic and covalent, and to think of these as being entirely different. But in fact these merely represent two extremes, and many other intermediate types are possible. We can illustrate this by considering hydrogen chloride. If we imagine the bonding in hydrogen chloride to be covalent, then the two bonding electrons are shared exactly equally between the two atoms; if we imagine it to be ionic, then both electrons reside entirely on the chlorine. But we may also imagine many intermediate situations in which the electrons are shared between the two atoms, but not equally, so that the electrons are on average closer to one atom than the other. In reality it is found that such an intermediate description is the most accurate, and the H–Cl bond is sometimes said to be covalent with some ionic character.

So far we have used the ionic model to describe transition metal compounds, and this is reasonable for cases where the charge on the metal ion is low. But as the charge on the metal ion increases, the ion becomes better at attracting electrons to itself, and so the degree of covalency in its chemical bonding increases. Thus it might be thought that CrO_3 ought to contain Cr^{6+} ions, but in reality the bonding between the Cr and O atoms is at least partially covalent. That is not to say that CrO_3 is a covalent compound, and that the charge on the Cr is really zero; it is simply that because the bonding is not entirely ionic, the plus charge on the Cr is not as high as 6.

Nevertheless we would expect the charge on the metal atom in CrO_3 to be greater than that in CrO, and it is often convenient to use the concept of oxidation number to reflect this. The oxidation number of an element may be defined as the charge that its atom would carry in a compound if the bonding were completely ionic; thus in CrO_3 we say that the oxidation number of Cr is +6, without implying anything about the nature of the bonding. This turns out to be useful for two reasons. Firstly, it helps in balancing equations in redox reactions, being a simple method of ensuring that the total number of electrons does not alter during a reaction. Secondly, it helps group together compounds with similar properties; thus $K_2Cr_2O_7$ and CrO_3 do resemble each other chemically, even if neither actually contains Cr^{6+} ions.

General patterns. Table 5.3 shows some of the more important oxidation states found in compounds of the elements of the first transition series. Although the range is clearly very wide, there are some general trends which can be seen.

For the elements in the first half of the transition series there exist compounds in which the metals have at least formally lost all their valence electrons. Thus we have Sc_2O_3, TiO_2, V_2O_5, CrO_3 and Mn_2O_7. As we move across the transition series it becomes increasingly difficult to remove all the valence electrons, mainly because of the large positive charge left on the metal ion. Figure 5.2 shows the total energy required to remove all the 4*s* and 3*d* electrons from the elements Sc to Fe. For scandium the energy is appreciably less than that required to form Al^{3+} from Al, and scandium forms 3+ ions in virtually all cases, whereas Mn_2O_7 may be formed only under strongly

Table 5.3 *Common oxidation states in the first transition series*

Oxidation state	Sc	Ti	V	Cr	Mn	Fe	Co	Ni	Cu
+7					Mn				
+6				Cr					
+5			V						
+4		Ti	V		Mn				
+3	Sc	Ti	V	Cr		Fe			
+2		Ti	V	Cr	Mn	Fe	Co	Ni	Cu
+1									Cu

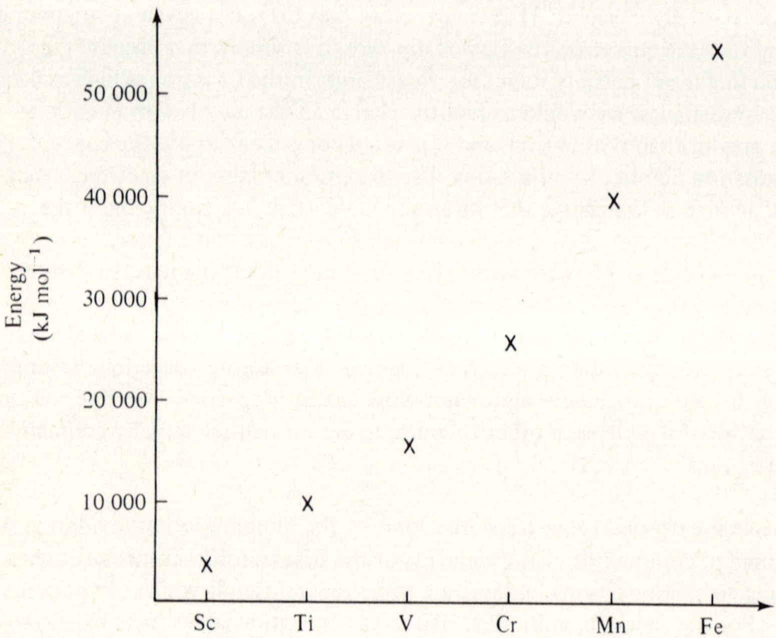

5.2 *Energy required to remove all the valence electrons from the elements Sc to Fe*

oxidising conditions. The converse of this is that the oxidising nature of these states increases across the series; Sc_2O_3 has no oxidising properties, whereas Cr(VI) and Mn(VII) are very strong oxidising agents and are commonly used as dichromate ($Cr_2O_7^{2-}$) and permanganate (MnO_4^-). They are strongly oxidising as they attract electrons to themselves very powerfully. By the time that iron is reached, the energy required to remove all eight electrons is enormous, and is never recouped from the lattice energy or partial covalency. Fe forms a few compounds such as $BaFeO_4$, in which the oxidation number is +6, but its most important compounds have oxidation numbers +2 and +3.

The second general trend which may be seen from table 5.3 is the existence of a stable oxidation state +2 across the transition series. In V and Cr the +2 state is strongly reducing, and only stable in the absence of air, but it becomes more stable in the later elements, and by Co, Ni and Cu it is much the most important state. At first sight this might appear rather surprising; we are accustomed to oxidation numbers increasing or decreasing by one as we move across the periodic table. But this is simply the result of the traditional emphasis on the main group elements; in the transition metals, and also in the lanthanides and actinides, it is much more common to see one oxidation state recurring throughout a series of adjacent elements.

We can see why this should be so by considering the variation of two important physical properties across the periodic table, ionisation energies and ionic radii. These are shown in table 5.4. The ionic radius (r) is of significance because it is the main factor determining the lattice energy; the smaller the ionic radius, the larger the lattice energy, as the interaction energy of two ions is inversely proportional to the distance between their nuclei. Table 5.4 therefore gives us information on the two main factors which influence the heats of formation, and hence the stabilities, of the transition metal compounds MX_2.

Table 5.4 *Properties of the +2 ions in the first transition series*

	Sc	*Ti*	*V*	*Cr*	*Mn*	*Fe*	*Co*	*Ni*	*Cu*
1st E_i (kJ mol^{-1})	630	660	650	650	720	760	760	740	750
2nd E_i (kJ mol^{-1})	1200	1300	1400	1600	1500	1600	1600	1800	2000
$r(M^{2+})$ (nm)	–	0.090	0.088	0.088	0.080	0.076	0.074	0.072	0.069

The most striking feature of the table is the rather small variation in both ionisation energies and ionic radii across the transition series. The ionisation energies show little change as we are considering the removal of electrons from the same orbital in each case; the ionic radii are similar as they are all determined mainly by the size of the partially-filled *d* shell. There is a slow increase in ionisation energy, which reflects the increasing nuclear charge, binding the valence electrons rather more strongly. There is also a decrease in ionic radius, again caused by the increasing nuclear charge; it therefore follows that lattice energies will increase slowly across the period, to some extent cancelling out the slow increase in ionisation energies. We can see that if we were to construct a series of Born–Haber cycles for the various transition metals in their various oxidation states, there would be considerable similarities between the cycles for the different elements, and so conclusions about the stability of a particular oxidation state might well be the same for several different elements. We would thus expect one state to feature strongly in the chemistries of a number of adjacent elements.

Notice that this argument does not specifically predict that the +2 state will be the most important in the first transition series; we should require a much more detailed analysis of the relative magnitudes of various terms before we could draw that conclusion. However the argument is not restricted to the first transition series, and similar patterns can be found elsewhere in the periodic table. Thus in the lanthanides the +3 state is very stable for the majority of elements, and the + 4 state is important for the elements in the third transition series from Hf to Au.

Half-filled and filled *d* shells. It is often said that in transition metals there is an extra stability associated with an atom or ion which possesses either a half-filled or completely-filled *d* shell. It is much less common to find any explanation of this alleged extra stability, and one is tempted to conclude that there is something inherently desirable in electron shells which are either full or exactly half-full. This is strangely reminiscent of elementary treatments of valence theory, in which main group elements gain and lose electrons so as to acquire complete electron shells. As we saw in chapter 4, this is a useful rule-of-thumb which predicts many chemical formulae correctly, but it does not answer the question of why complete electron shells are formed in the first place. A more detailed study of ionisation and lattice energies is needed to solve that problem, and in a similar way we must try to explain why half-filled and filled *d* shells seem to be particularly stable.

Several pieces of evidence are advanced to show that half-filled and filled *d* shells are unusually favourable. As we saw in table 5.1, the electron configurations of the elements of the first transition series are generally of the type $4s^2 3d^n$. However there are two exceptions to this rule, Cr with the configuration $4s^1 3d^5$ and Cu with the configuration $4s^1 3d^{10}$. These elements would be expected to have the configurations $4s^2 3d^4$ and $4s^2 3d^9$ respectively, and it is suggested that their anomalous behaviour is caused by the extra stability given by their half-filled and filled *d* shells.

Copper compounds provide further support for the theory: copper is the only element in the first transition series to form 1+ ions. As Cu^+ ions have the electron configuration $3d^{10}$, it is suggested that their unique stability may derive from their completely-filled *d* shell.

The difference in ease with which Mn^{2+} and Fe^{2+} may be oxidised to the corresponding +3 ions has also been attributed to the influence of half-filled electron shells. It is well known that Fe^{2+} ions are readily oxidised to Fe^{3+}, whereas Mn^{2+} ions are more difficult to oxidise and this is confirmed by their redox potentials (Fe^{2+}/Fe^{3+} +0.77V, Mn^{2+}/Mn^{3+} +1.15V). In the oxidation of

Fe the electron configuration goes from $3d^6$ to $3d^5$, and the favourability of the final half-filled shell is said to explain the ease with which oxidation occurs. In contrast the oxidation of manganese involves a change from $3d^5$ to $3d^4$, in which the stability of the half-filled shell is lost: Mn^{2+} is therefore more difficult to oxidise to Mn^{3+}.

However it is dangerous to assign too great a significance to these few cases. There are many other examples in transition metal chemistry where the supposed influence of half-filled or filled shells is more difficult to detect: for example the ions Cr^+ and Co^{4+}, which both have the configuration $3d^5$, are very unstable, and the electron configurations of the atoms of the second transition series do not show the same neat pattern as those of the first transition series. There is certainly no inherent desirability in electron shells being either half- or completely-filled, and we must seek a rational explanation of why the configurations d^5 and d^{10} occur rather more commonly than might at first be expected.

The simplest measure of the stability of any electron is its ionisation energy; if we are concerned with the stabilities of *d* electrons in transition metals, then we must consider their third ionisation energies, as the first and second ionisation energies usually refer to the removal of a 4*s* electron.

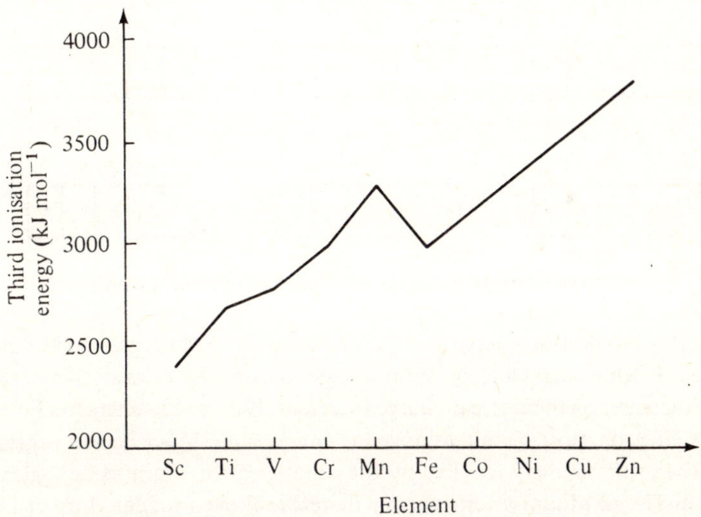

5.3 Third ionisation energies of the elements of the first transition series

Figure 5.3 shows the variation in third ionisation energy across the elements Sc to Zn. The stabilities of electrons in atoms are generally determined by two main factors, the attraction of the positively-charged nucleus and the repulsion of the other negatively-charged electrons. The main trend shown in figure 5.3 is an increase in ionisation energy across the transition series, and this is caused by the increasing positive charge on the nucleus holding the *d* electrons more strongly. But superimposed on this general trend is a discontinuity at Fe, and this is caused by a sudden change in electron repulsion. In a transition metal ion there are five *d* orbitals, and they all have exactly the same energy; each orbital can hold a maximum of two electrons, making ten in all. In the ion Sc^{2+} there is just one *d* electron, and so it does not matter which orbital it occupies. The ion Ti^{2+} has two *d* electrons, and so there are now two possibilities: both electrons could occupy the same orbital, or they could occupy two different orbitals. It is found experimentally that the second alternative is the more stable; the electron–nuclear attraction is the same in each case, but the repulsion between the two electrons is lower if they occupy different regions of space, rather than the same region. In general, when there is a choice of orbitals all with the same energy, electrons occupy as many different orbitals as possible: this is known as Hund's rule. But when the *d* shell is more than half-full, then of course some orbitals have to contain two electrons, as is shown in figure 5.4.

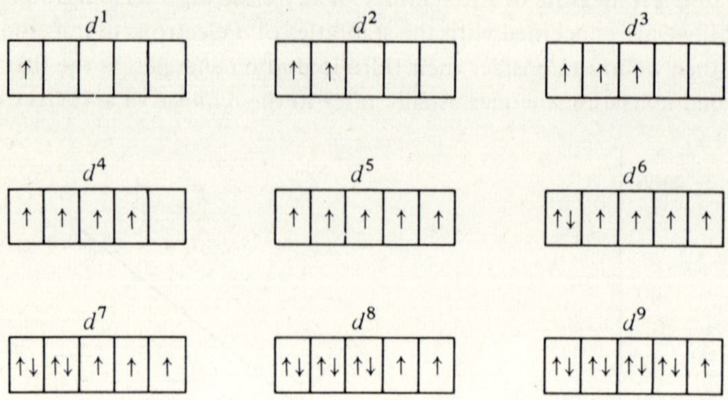

5.4 *Distributions of* d *electrons in partially-filled* d *shells*

The third ionisation energies of the elements Sc to Mn represent the energy required to remove an electron from a singly-occupied *d* orbital; this energy slowly increases as the nuclear charge increases. But for the elements Fe to Zn, the electron most readily removed is now in a doubly-occupied orbital, and is therefore strongly repelled by the other electron in the same region of space. The third ionisation energies therefore show a sudden drop at Fe, and then continue to rise smoothly again as the nuclear charge continues to increase. There is then a further drop after Zn, when the third ionisation energy then corresponds to the removal of a 4*s* and not a 3*d* electron.

We can therefore see that the stabilities of 3*d* electrons increase steadily across the transition series, with a break when the *d* shell becomes half-full. This is the proper explanation of why half-filled and filled *d* shells are sometimes thought to have some special stability. If we consider removing an electron from Mn^{2+} and Fe^{2+}, then we should expect the Fe^{2+} to be more readily oxidised, as we are removing an electron from a doubly-occupied orbital, whereas in Mn^{2+} we are removing an electron from a singly-occupied orbital. Similarly if +1 ions are to exist anywhere in the transition series, we would expect the most likely point to be on the right-hand side, at Cu where the *d* electrons are at their most stable. If neutral atoms are to have the configuration $4s^1 3d^{n+1}$ rather than $4s^2 3d^n$, we would expect this to be most likely at Cr and Cu, where the *d* electrons are most stable.

If we wish to remember this effect as being a feature of half-filled and filled *d* shells, that may well be a useful *aide-mémoire*; but we should not forget that the explanation of the effect lies in the way that electron repulsion varies across the transition series.

C Catalysis

The phenomenon of catalysis of chemical reactions is well known: a catalyst may be defined as a substance which alters the rate of a chemical reaction, but which itself remains chemically unaltered at the end of the reaction. Experimentally it is found that the transition metals and their compounds are capable of catalysing a wide range of chemical processes: table 5.5 lists a few examples which are of practical importance.

Table 5.5 *Reactions catalysed by transition metals*

Ni	H_2 + alkenes = alkanes	
V_2O_5	$2SO_2 + O_2 = 2SO_3$	Contact process
CuO	$4HCl + O_2 = 2Cl_2 + 2H_2O$	Deacon process
Fe_2O_3	$N_2 + 3H_2 = 2NH_3$	Haber process
$AlCl_3/Ti(C_2H_5)_4$	Polymerisation of alkenes	Ziegler–Natta

Although a wide range of substances can act as catalysts in many different reactions, all catalysts operate in essentially the same way. They do not alter the position of equilibrium of a reaction in any way, but simply alter the rate at which equilibrium is reached by providing an alternative reaction pathway which is quicker. This alternative pathway has a lower activation energy than the uncatalysed reaction. The rate of a reaction is related to the activation energy E_a by the equation

$$\text{Rate} = A \exp(-E_a/RT)$$

where R is the gas constant and T is the temperature. A is a constant, called the frequency factor. The activation energy may be determined experimentally from the variation of reaction rate with temperature, but we may think of it in simple terms as representing the minimum energy required by two colliding molecules for reaction to occur. The action of a catalyst is often represented by a diagram such as figure 5.5; notice that the overall energy change for the reaction is unaffected by the catalyst. The precise shape of

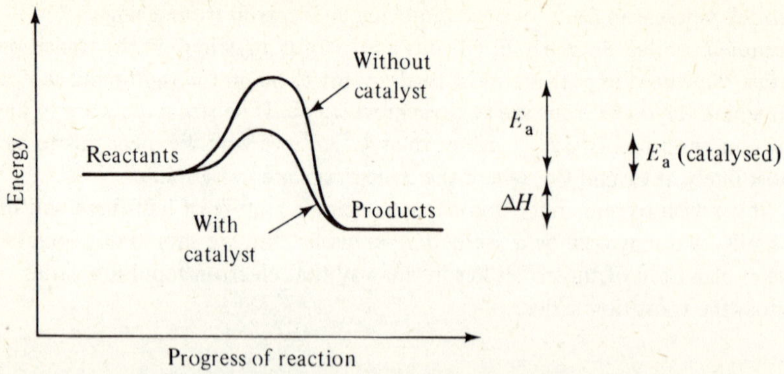

5.5 *The effect of a catalyst on activation energy*

the two curves is not important, and in some cases they may show more than one maximum. The catalysed reaction always has a lower activation energy, however, and therefore proceeds more rapidly. (The question of whether A remains the same when the catalyst is added is often conveniently ignored; in fact it can change, but the change caused by the lowering of E_a is always much greater.)

Many reactions catalysed by transition metals are of considerable industrial or synthetic importance, and they have been studied in some detail. As a result the precise way in which individual catalysts operate is often well understood; unfortunately it is not a simple matter to go on to predict which other substances might act as catalysts, for it is quite common for a reaction to be catalysed by one substance but not by another which appears to be chemically similar. The development of new catalysts still involves a strong element of trial and error.

One of the main reasons why transition metals and their compounds find such wide use as catalysts is their ability to form several stable oxidation states; however we may be wary of believing that this can explain every example we encounter, for other properties such as the ability to form complex ions can also play an important role.

Homogeneous catalysis. When a reaction occurs with all the reagents in the same phase, it is called homogeneous; if two or more phases are present, for example when a solid is in contact with a gas, it is called heterogeneous. Transition metal ions often act as homogeneous catalysts in solution, while both transition metals and their solid compounds can act as heterogeneous catalysts.

One of the commonest type of reactions catalysed by transition metal ions are redox reactions. A simple example is the reaction of Ce^{4+} with Tl^{+}:

$$2Ce^{4+} + Tl^{+} = 2Ce^{3+} + Tl^{3+}$$

This reaction is catalysed by Mn^{2+} ions, and it is thought that the catalyst works in the following way. The Mn^{2+} is oxidised by the Ce^{4+} first to Mn^{3+} and then to Mn^{4+}:

$$Ce^{4+} + Mn^{2+} = Ce^{3+} + Mn^{3+}$$
$$Ce^{4+} + Mn^{3+} = Ce^{3+} + Mn^{4+}$$

The Mn^{4+} then oxidises the Tl^{+} to Tl^{3+}, which regenerates Mn^{2+} ions:

$$Mn^{4+} + Tl^{+} = Mn^{2+} + Tl^{3+}$$

The cycle can then begin again; clearly the important feature of the catalyst here is its ability to be oxidised and reduced readily.

A similar example is the use of transition metals to catalyse the reaction of $S_2O_8^{2-}$ with I^-. This reaction is slow at room temperature, but its rate increases markedly on the addition of Fe^{2+}, Fe^{3+} or Cu^{2+} ions. The mechanism is:

$$2I^- + 2Fe^{3+} = I_2 + 2Fe^{2+}$$
$$\underline{2Fe^{2+} + S_2O_8^{2-} = 2Fe^{3+} + 2SO_4^{2-}}$$
$$2I^- + S_2O_8^{2-} = I_2 + 2SO_4^{2-}$$

Copper acts similarly, using the oxidation states Cu(II) and Cu(I). This reaction illustrates some of the difficulties involved in trying to predict other possible catalysts. It might be thought that any transition metal with two oxidation states, the lower of which could be oxidised by $S_2O_8^{2-}$, and the higher of which could oxidise I^-, could catalyse this reaction. This condition proves to be necessary, but not sufficient: Cr(III) can be oxidised to Cr(VI) by $S_2O_8^{2-}$, and Cr(VI) can oxidise I^- returning to Cr(III) but this alternative pathway proves to be no faster than the uncatalysed reaction between $S_2O_8^{2-}$ and I^-. The intervention of the chromium does not lower the activation energy, and so no catalysis is observed.

Transition metal ions can also catalyse the reactions of covalent molecules. Here they act as one-electron oxidising and reducing agents, forming free radicals. An example of this type of reaction is the oxidation of ethanol by hydrogen peroxide, which is catalysed by Fe^{2+} ions. The first step is the oxidation of Fe(II) to Fe(III) by H_2O_2:

$$Fe^{2+} + H_2O_2 = Fe(OH)^{2+} + \cdot OH$$

The reactive $\cdot OH$ radical then abstracts a hydrogen atom from ethanol:

$$\cdot OH + CH_3CH_2OH = H_2O + CH_3\dot{C}HOH$$

The resulting radical reduces Fe(III) to Fe(II) again:

$$CH_3\dot{C}HOH + Fe(OH)^{2+} = Fe^{2+} + H_2O + CH_3CHO$$

The examples we have seen so far all rely on the ability of transition metal ions to undergo ready redox reactions. But not all catalytic activity depends on this property; the ability to form complex ions can also be important. This may assist in the formation of reactive species, or may simply bring reactants into close proximity. For example the reaction of bromine with benzene is catalysed by $FeBr_3$; the mechanism of this reaction may be crudely represented:

$$FeBr_3 + Br_2 = FeBr_4^- + Br^+$$
$$Br^+ + C_6H_6 = C_6H_5Br + H^+$$
$$H^+ + FeBr_4^- = HBr + FeBr_3$$

Here the function of the $FeBr_3$ is to assist in the formation of Br^+ ions, and the oxidation number of the iron does not change during the reaction.

Heterogeneous catalysis. Transition metals and their compounds find wide use as heterogeneous catalysts: nickel is used in the hydrogenation of alkenes, and vanadium(V) oxide is used in the production of sulphur trioxide. When heterogeneous catalysis occurs reactant molecules, originally in either the gas or liquid phase, combine while adsorbed on the surface of the catalyst. The product is then desorbed from the catalyst surface, leaving space for further reactant molecules.

The key step in heterogeneous catalysis is the adsorption of molecules onto the solid surface. This adsorption is called chemisorption, as it involves the formation of relatively strong bonds between a molecule and the surface atoms of the catalyst. The precise nature of species adsorbed on solid surfaces is an active area of research, but it is important to realise that the strong bonds to the catalyst may modify the properties of reactant molecules considerably. For example when hydrogen molecules are adsorbed on metal surfaces it is sometimes said that hydrogen atoms are formed; but these hydrogen atoms do not behave as they would in the gas or liquid phases. It seems doubtful whether any of the particles adsorbed on solid surfaces may usefully be described as odd-electron species.

D Colour

One of the best known and most useful properties of the transition metals is the colour of their compounds: the deep blue of the copper–ammonia complex, the blood-red obtained by the addition of thiocyanate to iron(III), and the use of the purple colouration of permanganate as an indicator in titration are familiar. Although colour is not restricted to transition metal compounds (the yellow of sulphur and the red of diazo-dyes are simple exceptions) it is not a common property of any other group of elements. Not surprisingly the specific feature of transition metals which gives rise to their colour is their partially-filled *d* shells.

It is important not to confuse the origins of the colours of a sodium lamp, which we discussed in chapter 4, and of a transition metal compound such as copper sulphate. A sodium lamp appears yellow even in a darkened room, whereas a bottle of copper sulphate solution appears blue only when it is illuminated by white light; in a darkened room copper sulphate is invisible. A sodium lamp emits yellow light; transition metal compounds appear coloured because they absorb some of the colours of white light, and transmit others. Thus copper sulphate appears blue because it does *not* absorb blue light, but does absorb red and yellow light.

The processes of light emission and absorption are really complementary. When visible or ultra-violet light is emitted, an electron originally in an excited orbital falls to a more stable orbital, and gives out a photon. When visible or ultra-violet light is absorbed, an electron originally in its most stable orbital jumps to a higher orbital, using the energy absorbed from the incoming photon. An ion will therefore absorb photons whose energy happens to coincide with the energy required to promote an electron from its ground state to another orbital. For a potassium ion, which has 18 electrons, a simple transition might be:

$$1s^2\ 2s^2\ 2p^6\ 3s^2\ 3p^6 \longrightarrow 1s^2\ 2s^2\ 2p^6\ 3s^2\ 3p^5\ 4s^1$$

If the energy required to promote an electron corresponds to a photon in the visible region, then the ion will appear coloured. As we saw on p. 13, this requires the energy to fall in the rather narrow range 3 to 5×10^{-19} J.

However for the great majority of stable ions the amount of energy required to promote an electron from its ground state even to the lowest excited orbital is greater than 5×10^{-19} J, and these ions can therefore only absorb ultra-violet light. They transmit all visible light, and appear colourless to the human eye. Even for transition metal ions, the promotion of a $3d$ electron to a $4p$ orbital requires more energy than is available from visible photons, and we shall need to look elsewhere for an explanation of the colours of transition metal ions.

Ligand field splitting. A few simple experiments on transition metal ions in solution can give an important clue as to the origin of their colours. Anhydrous copper sulphate is white; solid hydrated copper sulphate and its aqueous solution are light blue. When concentrated ammonia solution is added, copper sulphate solution turns deep blue; if excess concentrated hydrochloric acid is now added, the solution becomes light green.

The explanation of these facts is well known: copper forms complex ions in which the metal is surrounded by four ligands. These ligands may be neutral molecules such as water or ammonia, or negative ions such as chloride or EDTA. Solid hydrated copper sulphate and its solution both contain the ion $Cu(H_2O)_4^{2+}$, which is light blue in colour. When ammonia is added the water molecules are replaced by a stronger ligand, and the deep blue ion $Cu(NH_3)_4^{2+}$ is formed. When chloride ions are added another change of ligand occurs, and $CuCl_4^{2-}$ (a green ion) is produced.

We may think of the bonding between the transition metal ion and the ligands as being electrostatic in origin. The water molecules in $Cu(H_2O)_4^{2+}$ are oriented so that the negative end of the molecule (the oxygen atom) is closest to the metal, whereas the positive end (the hydrogen atoms) is further away. This results in a net attraction between the metal ion and the dipolar ligand molecule. In $CuCl_4^{2-}$ the attractive force follows directly from the opposite charges of the metal ion and ligands. This electrostatic model is something of an oversimplification as dative covalency may also be involved, but it will prove quite sufficient for our purposes.

All the transition metals form complex ions, and the colours of these ions depend on the natures of both the metal and the ligand. The complex ions usually contain either four or six ligands, although other coordination numbers are occasionally found.

Clearly the ligands play an important role in determining the colour of transition metal compounds. But the mere formation of complex ions is not sufficient to produce colour; elements such as sodium and magnesium also form complex ions, but these are colourless. The colour of transition metal ions derives from the interaction of the ligands with a partially-filled *d* shell; to understand the nature of this interaction we shall need to look closely at the shapes of *d* orbitals.

Quantum mechanics tells us that we must abandon any attempt to describe the precise path of an electron in an atom; instead we may represent an electron by a sort of cloud, whose density gives the probability of finding the electron at any particular point. The simplest shape is that of an *s* orbital; it is spherical, and may be drawn as in figure 5.6. However

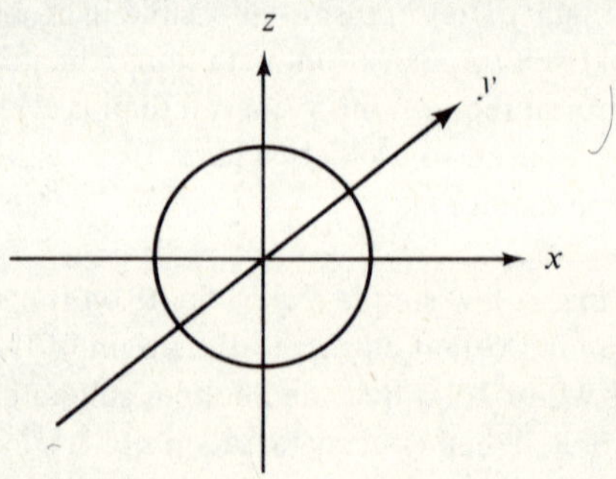

5.6 *1s orbital*

orbitals must not be thought of as having definite edges beyond which the electron may not be found; the cloud density simply becomes smaller and smaller further away from the nucleus. 1*s*, 2*s* and 3*s* orbitals all have the same shape, but their points of highest electron density are increasingly further away from the nucleus.

An atom has three 2*p* orbitals; each can hold a maximum of two electrons, making a total of six electrons when all three orbitals are full. They each have exactly the same shape and size, and therefore the same energy, but they point in different directions in space, as shown in figure 5.7. 3*p* and 4*p* orbitals are similar, but are of course rather larger.

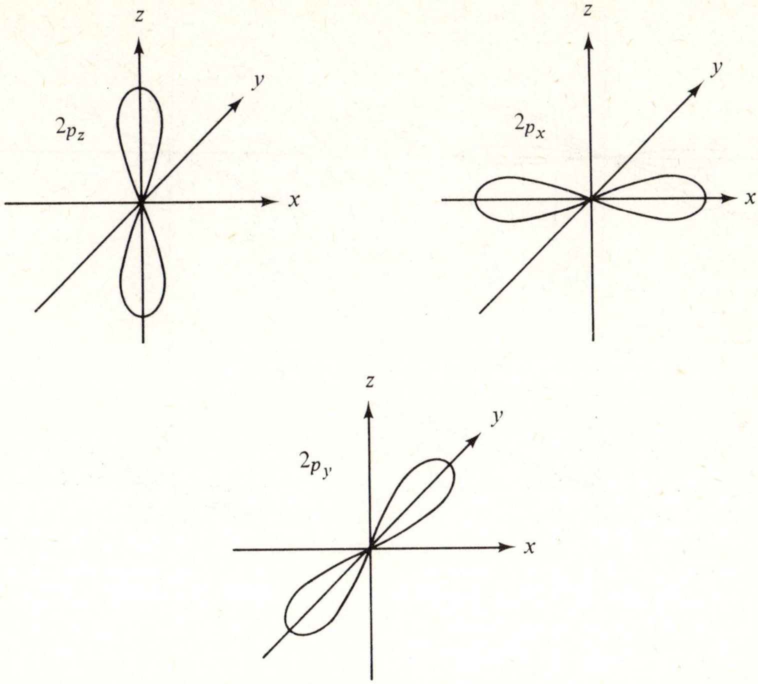

5.7 2p *orbitals*

There are five 3*d* orbitals in an atom; as each can hold up to two electrons, the maximum number of 3*d* electrons per atom is 10. Four of these orbitals have the same shape as each other; the fifth is not similar to the others but nevertheless they all have exactly the same energy. They are shown, together with their labels, in figure 5.8.

We must now consider how the *d* orbitals are affected by the presence of the ligands in a complex ion. If the complex ion contains six ligands, then it is found that they are arranged at the corners of a regular octahedron, as shown in figure 5.9. When an ion is surrounded by four ligands, then a tetrahedral arrangement is most common; however in a few cases the ligands are arranged in a square, with the metal ion at the centre. Figure 5.10 shows the structure of the tetrahedral ion $CuCl_4^{2-}$.

We may now show why the presence of the ligands does not affect all five *d* orbitals equally. For simplicity we shall consider only two of the *d* orbitals, and we shall select two in the same plane. We shall consider an octahedral arrangement of ligands, of which only the four in the plane of the paper will be shown. There are actually two more ligands, one above and one below the plane of the paper, but these do not affect the argument in any way. Figure 5.11 shows the two *d* orbitals in separate diagrams for clarity: in reality one diagram would be superimposed on the other. The ligands are labelled L. In the absence of the ligands, the two *d* orbitals have

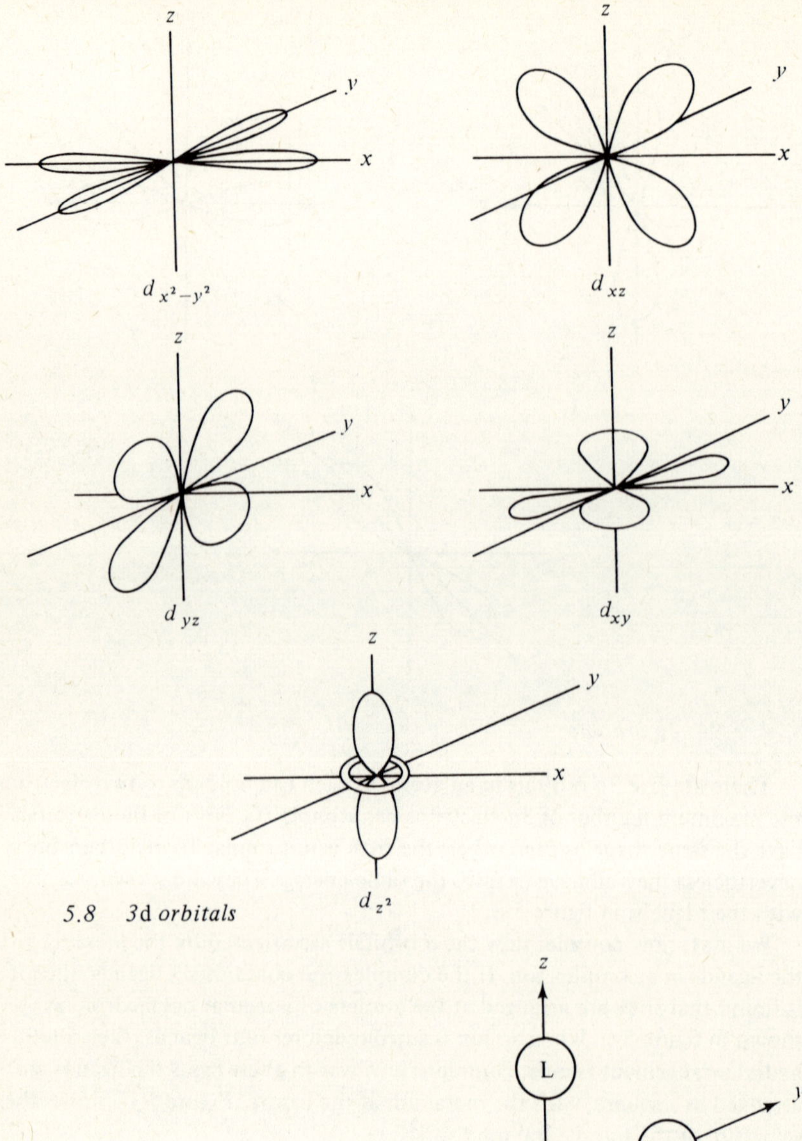

5.8 3d *orbitals*

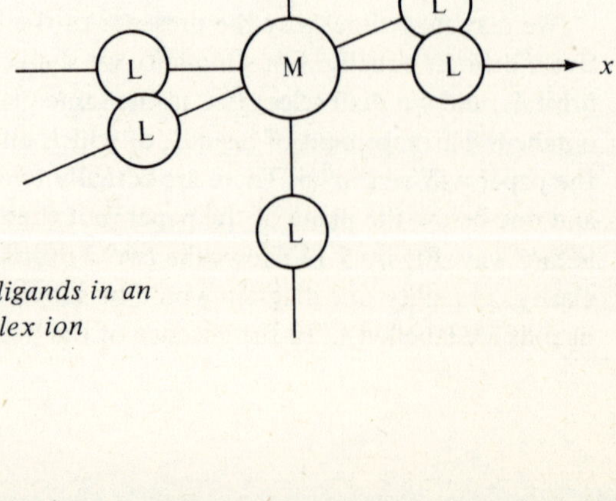

5.9 *Arrangement of ligands in an octahedral complex ion*

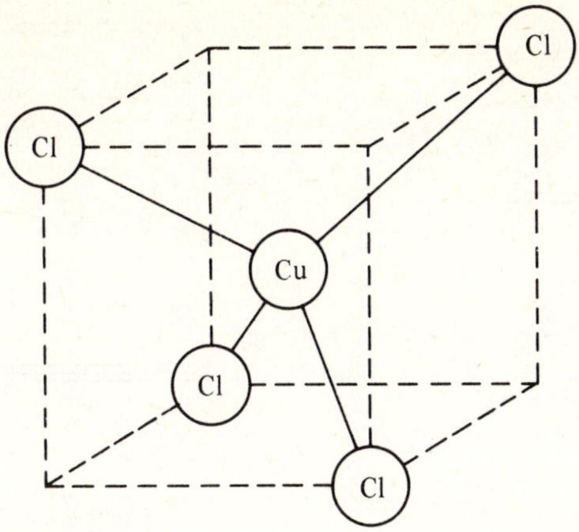

5.10 *The ion* $CuCl_4^{2-}$

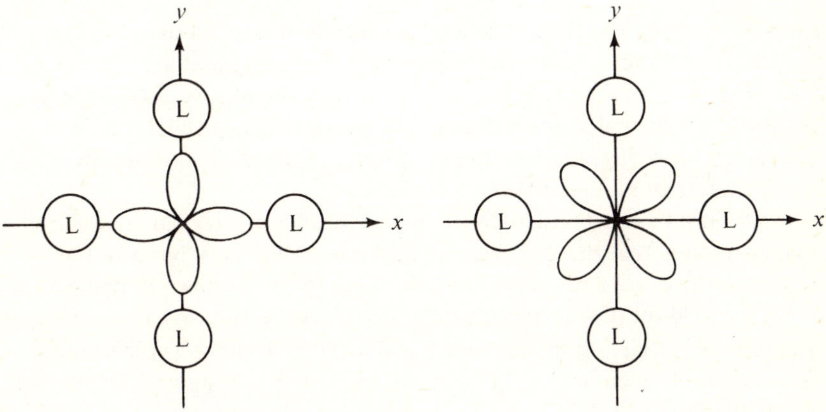

5.11 *The interactions of the* $d_{(x^2-y^2)}$ *and* d_{xy} *orbitals with an octahedral ligand field*

exactly the same energies, as they have identical shapes and sizes. But from figure 5.11 we can see that the ligands, which are negative ions or the negative ends of dipolar molecules, do not interact with them equally. In the left-hand diagram, the electron in the *d* orbital would spend much of its time very close to the ligands, which would repel it electrostatically. In the right-hand diagram, the electron spends most of its time rather further away from the ligands, and so would be repelled rather less. We would therefore expect that an electron in the right-hand orbital would be more stable than an electron in the left-hand orbital.

When this analysis is extended to include all five orbitals it is found that the ligands have the effect of splitting the *d* orbitals into two groups. The d_{xy}, d_{yz} and d_{xz} orbitals all have the same energies, and they are more stable than the $d_{(x^2-y^2)}$ and d_{z^2} orbitals which have the same energies as each other. As this

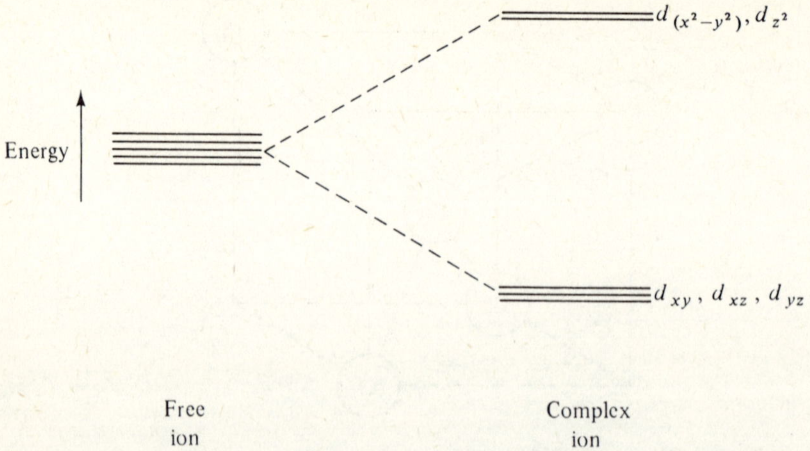

5.12 *Splitting of* d *orbitals in an octahedral complex*

effect is caused by the electric field of the ligands, it is called ligand field splitting, and is often represented as in figure 5.12. Such diagrams are convenient ways of showing the splitting, but they show only the relative energies of the *d* orbitals, not their absolute energies. Any electron in a *d* orbital will be repelled by the six ligands, but those in the d_{xy}, d_{yz} and d_{xz} orbitals are repelled less than those in the other orbitals.

If a transition metal is surrounded by four ligands in a regular tetrahedron, it is again found that the *d* orbitals are split into two groups; but now the $d_{(x^2-y^2)}$ and d_{z^2} are more stable than the other three, as shown in figure 5.13.

Ligand field splitting is important because the difference in energies between the *d* orbitals often falls in the range 3 to 5 $\times 10^{-19}$ J, and so a visible photon can excite an electron from one *d* orbital to another. For example a Ti^{3+} ion has the configuration [Ar] $3d^1$, and so in an octahedral complex such as $Ti(H_2O)_6^{3+}$,

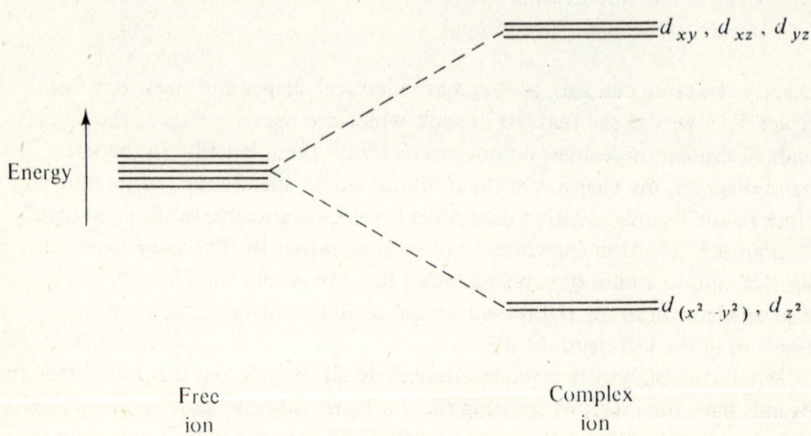

5.13 *Splitting of* d *orbitals in a tetrahedral complex*

it has one electron in one of the orbitals $3d_{xy}$, $3d_{yz}$ or $3d_{xz}$ (which one is immaterial). If the ion is irradiated with visible light, the electron can be promoted to the $3d_{(x^2-y^2)}$ or $3d_{z^2}$ orbital, as shown in figure 5.14. Similarly a Cu^{2+} ion has the electron configuration [Ar] $3d^9$; figure 5.15 shows the electron excitation which occurs when a tetrahedral complex such as $CuCl_4^{2-}$ absorbs a photon in the visible region.

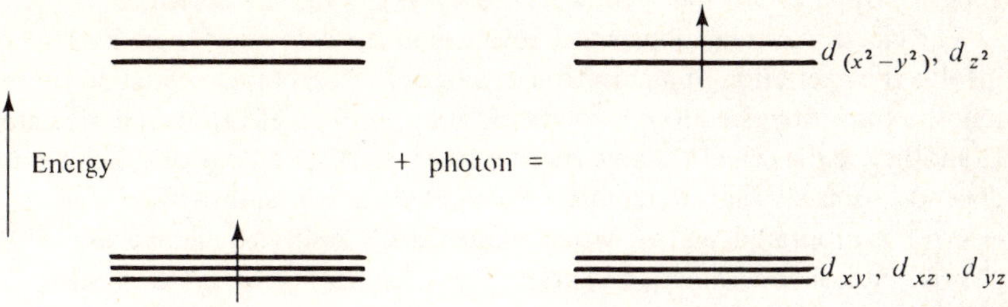

5.14 *Absorption of visible light by* $Ti(H_2O)_6^{3+}$

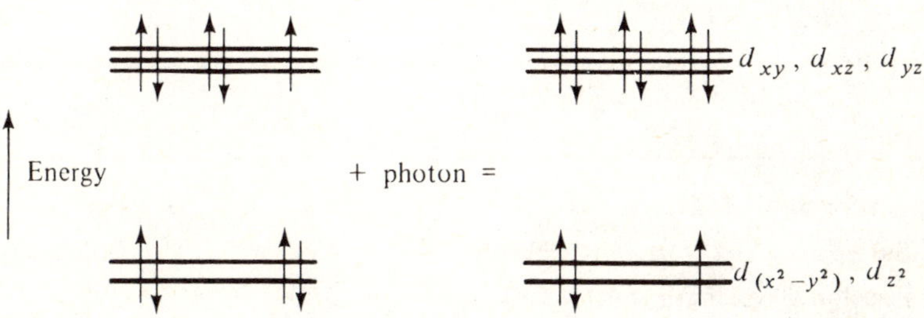

5.15 *Absorption of visible light by* $CuCl_4^{2-}$

Of course not all ligands produce exactly the same ligand field splitting in a transition metal ion; it is found that the CN^- ion produces very large splittings whereas the halide ions have a rather smaller effect. Thus in a series of complex ions with the same transition metal, we find that each ligand produces a slightly different splitting, and so each ion has a slightly different colour. Similarly the magnitude of the ligand field splitting depends also on the nature of the transition metal ion, and so a series of ions with the same ligands but different metals also differ in colour.

This theory also explains readily why only ions with partially-filled *d* shells are coloured. An ion such as Sc^{3+} contains no *d* electrons at all, and so although the *d* orbitals will still be split by the presence of ligands, it is not possible to excite an electron from one *d* orbital to another; thus Sc^{3+} ions do not absorb visible light at all, and appear colourless. Ions such as Zn^{2+} and Ag^+ contain full *d* shells, and again it is not possible to excite an electron from one *d* orbital to another; there is no vacancy in the less stable orbitals into which one of the more stable electrons could be promoted, and so Zn^{2+} and Ag^+ ions are also colourless.

Charge-transfer spectra. Ligand field splitting provides a satisfactory explanation of the colours of most transition metal ions, but it cannot be applied to every case. The ions VO_4^{3-}, CrO_4^{2-} and MnO_4^- form a series which consist at least formally of four oxide ions surrounding a V^{5+}, Cr^{6+} or Mn^{7+} ion. Each of these metal ions has no *3d* electrons, and we should therefore expect the complexes to be colourless. But experimentally it is found that MnO_4^- ions appear purple, CrO_4^{2-} ions yellow, and only VO_4^{3-} ions are colourless.

The electronic transitions which give rise to the colours of these ions involve the excitation of an electron in a ligand orbital to an orbital on the metal ion; the transition therefore involves the transfer of an electron from one atom to another, and is called a charge-transfer transition. The energy of this transition depends primarily on the stability of the metal *3d* orbital into which the electron is promoted, and we would expect this stability to increase as the positive charge on the metal increases. This is confirmed by table 5.6; the MnO_4^- ion requires the least energy to promote an electron from the ligand to the metal, and so it absorbs light of the longest wavelength.

Table 5.6 *Charge-transfer transitions in the ions MnO_4^-, CrO_4^{2-} and VO_4^{3-}*

	MnO_4^-	CrO_4^{2-}	VO_4^{3-}
Oxidation state of M	+7	+6	+5
Colour	Purple	Yellow	Colourless
Absorption wavelength (nm)	530	390	290
Transition energy (J)	3.8×10^{-19}	5.1×10^{-19}	7.0×10^{-19}

Charge-transfer transitions are not restricted to ions which formally contain no *d* electrons, and in fact they occur in most transition metal complexes. But they are usually found at rather higher energies than the *d–d* transitions which we saw in the last section, and so generally give rise to absorption of ultra-violet rather than visible light. Charge-transfer transitions are of greater importance in determining colour in other parts of the periodic table; for example the colours of the halogens, and of B-metal compounds such as PbI_2, are caused by charge-transfer transitions.

E Paramagnetism

Another physical property which is typical of transition metal compounds is paramagnetism. When substances which are paramagnetic are placed in a magnetic field, they acquire a small magnetic moment which reinforces the field, and they are thereby attracted to areas where the magnetic field is strongest. They attract magnetic lines of force to themselves, and so have permeabilities greater than one.

The phenomenon of paramagnetism is caused by electron spin. All electrons spin on their own axis, in much the same way as the Earth spins while orbiting the sun; as they are electrically charged this gives them a magnetic moment, and so they act like tiny permanent magnets. In many compounds this is of no consequence whatsoever for, when any orbital is doubly occupied, the spin of one electron directly opposes the spin of the other, and so the two magnetic moments cancel each other out. The electrons are then said to be paired. All compounds which have complete electron shells thus have no magnetic moment and do not show paramagnetism. But ions and molecules which have unpaired electrons do have magnetic moments, and these can align themselves with an external magnetic field. The transition metals, including the lanthanides and actinides, form the largest group of such substances, although other odd-electron species such as NO and NO_2 also behave in the same way.

Clearly any transition metal ion with an odd number of electrons will be paramagnetic, as it must have at least one unpaired electron. But a transition metal ion with an even number of electrons will also be paramagnetic, provided that it possesses a partially-filled *d* shell. This is because of Hund's rule, which states that the most favourable way of distributing the electrons amongst the available orbitals is to have the maximum possible number of singly occupied orbitals, and for all the unpaired electrons to have their spins parallel. Table 5.7 shows the arrangements of the electrons for the ions of the first transition series. The number of unpaired electrons rises to a maximum at Mn^{2+} and Fe^{3+}; ions with no *d* electron or with a complete *d* shell are not paramagnetic at all. It can be shown theoretically that the magnetic moment of an ion with *n* unpaired electrons should be $\sqrt{(n(n+2))}$, and these values are given in table 5.7. Experimentally the magnetic moment of an ion can be measured by a Gouy balance, which measures the force on a sample placed in an inhomogeneous magnetic field; it is found that the experimental magnetic moments are in most cases very close to the calculated values.

Table 5.7 *Paramagnetism of ions in the first transition series*

No. of d *electrons*	*Examples*	*Configuration*					*Unpaired electrons*	*Calculated magnetic moment*
0	Sc^{3+} Ti^{4+}			–			0	0
1	Ti^{3+} V^{4+}	↑					1	1.7
2	V^{3+}	↑	↑				2	2.8
3	V^{2+} Cr^{3+}	↑	↑	↑			3	3.9
4	Mn^{3+}	↑	↑	↑	↑		4	4.9
5	Mn^{2+} Fe^{3+}	↑	↑	↑	↑	↑	5	5.9
6	Fe^{2+}	↑↓	↑	↑	↑	↑	4	4.9
7	Co^{2+}	↑↓	↑↓	↑	↑	↑	3	3.9
8	Ni^{2+}	↑↓	↑↓	↑↓	↑	↑	2	2.8
9	Cu^{2+}	↑↓	↑↓	↑↓	↑↓	↑	1	1.7
10	Zn^{2+} Cu^{+}	↑↓	↑↓	↑↓	↑↓	↑↓	0	0

6 Mass spectrometry

A Introduction

As we saw in chapter 1, one of the simplest methods of producing odd-electron species is to bombard a stable, neutral molecule with high-energy electrons. These electrons provide the energy required to remove an electron from the molecule, so that positive ions are produced. Notice that the bombarding electrons have far too much energy to allow them to 'stick' to the target molecules, and so there is no question of negative ions being formed. The odd-electron species which are formed are generally of low stability, and electron bombardment has no synthetic value, but it is of great importance as it forms the basis of mass spectrometry.

Mass spectrometry is essentially a method of determining the masses of individual molecules. If all the molecules in a given sample do not have the same mass, the technique gives the mass of each type of molecule present, and its relative abundance. In a mass spectrometer neutral molecules are converted into positive ions by electron bombardment, and the masses of these ions are then measured from the deflections caused by electric and magnetic fields. Figure 6.1 shows the main features of a simple mass spectrometer. Molecules diffuse from the sample chamber on the left of the diagram into the region of the electron gun. Here they collide with high-energy electrons forming positive ions:

$$\mathrm{M} + \mathrm{e}^- = \mathrm{M}^+ + 2\mathrm{e}^-$$

They then pass into an electric field, which accelerates them towards the right of the diagram; if the potential difference across this region is V, and the charge on the electron is e, then the kinetic energy gained by the ion is:

$$E_\mathrm{k} = eV$$

Representing the mass of the ion by m and its velocity by v:

$$\tfrac{1}{2}mv^2 = eV \qquad (1)$$

Note that the whole spectrometer is carefully evacuated to prevent the ions losing kinetic energy by collisions with other molecules. The ion then passes into a uniform magnetic field, which is perpendicular to the plane of the paper in figure 6.1. As the ion is a moving charged particle, it experiences a

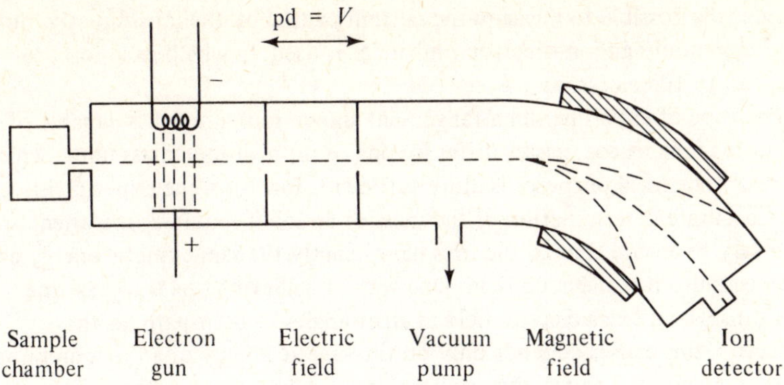

6.1 A simple mass spectrometer

force equal to Bev, where B is the magnetic flux density. (This effect of a magnetic field may be more familiar from the force exerted on a current-carrying wire, but a stream of ions constitutes an electric current just as much as electrons moving down a metal wire.) The ion is caused to move in a circular path: if the radius of the circle is r, then:

$$Bev = \frac{mv^2}{r}$$

Therefore $$r = \frac{mv}{Be}$$

But from equation **(1)**, we know that:

$$v = \left(\frac{2eV}{m}\right)^{1/2}$$

Therefore $$r = \left(\frac{2Vm}{B^2e}\right)^{1/2}$$

Therefore $$r^2 = \frac{2V}{B^2} \times \frac{m}{e}$$

We therefore see that if V and B are fixed, then the radius of the path of an ion is determined only by the ratio of its mass to charge; in this chapter we shall consider only experiments in which ions of charge +1 are formed, and so the path of an ion will depend on its mass only. Light ions have lower path radii than heavy ions, and are therefore deflected more by the magnetic field. A series of ion detectors suitably placed could count the number of ions corresponding to each mass, but in practice it is often more convenient to have just one detector, and to vary the magnetic field to bring ions of each mass to the detector in turn. In early experiments the darkening of photographic plates was used to detect ions; this is properly referred to as mass spectrography. Most modern apparatus uses electrometers, which measure charge. It is possible to measure the current caused by the ions directly, but more commonly an ion–electron multiplier is used, in which each ion produces up to 10^6 electrons.

The type of experimental arrangement shown in figure 6.1 is capable of separating ions whose masses differ by one or more atomic mass units, which for many chemical purposes is quite sufficient. But for some experiments a more accurate determination of the mass of an ion is required; it is then necessary to ensure that all the ions have exactly the same kinetic energy as they pass into the magnetic field. One way of achieving this is to pass the ions through an extra electric field at right-angles to their path, as the deflection this causes depends only on the kinetic energy of an ion, and not its mass. Figure 6.2 shows the arrangement in such a spectrometer. This instrument would be capable of measuring the mass of an ion to better than one part in 10^5.

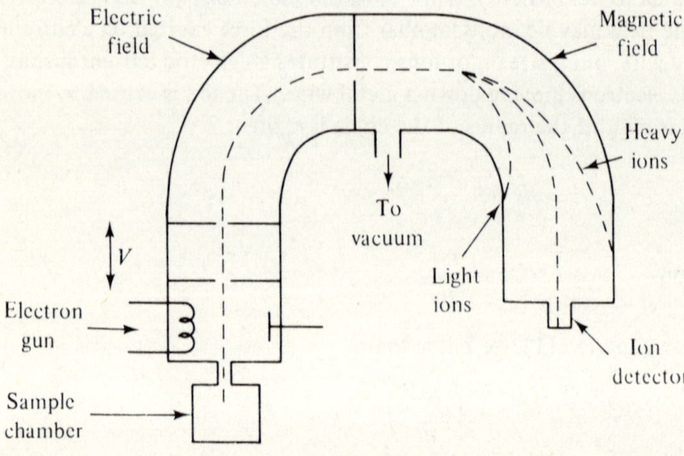

6.2 *Mass spectrometer for accurate work*

One unusual feature of mass spectrometry is the very small sample size required: most machines can detect less than 10^{-6} g of a substance. This is important as it allows the technique to be used on substances which are solids or liquids as well as on gases, provided that their vapour pressures are not vanishingly small. It also allows a mass spectrometer to be used in conjunction with other analytical techniques: for example the combination of chromatography and mass spectrometry is commonly used for the separation and identification of complex mixtures.

B Isotopes

One of the most remarkable features of a table of the relative atomic masses of the elements is that many of the values are very close to whole numbers; we are used to seeing H = 1, C = 12, N = 14 and O = 16. The explanation of this phenomenon is well known: atoms are composed of protons, neutrons and electrons, which have masses of roughly 1, 1 and 0 respectively. The mass of an atom is therefore equal to the number of protons plus neutrons, which must clearly be a whole number.

Unfortunately this explanation leaves unresolved the question of why some elements have masses which are not whole numbers; one such element is neon, which has a relative atomic mass of 20.18, and was one of the first elements to be studied by mass spectrometry. The mass spectrum of neon is shown in figure 6.3; it shows the masses of the ions produced and their relative abundances. We see that no neon atoms actually have mass 20.18, but that there are three different types of neon atom, with masses 20, 21 and 22.

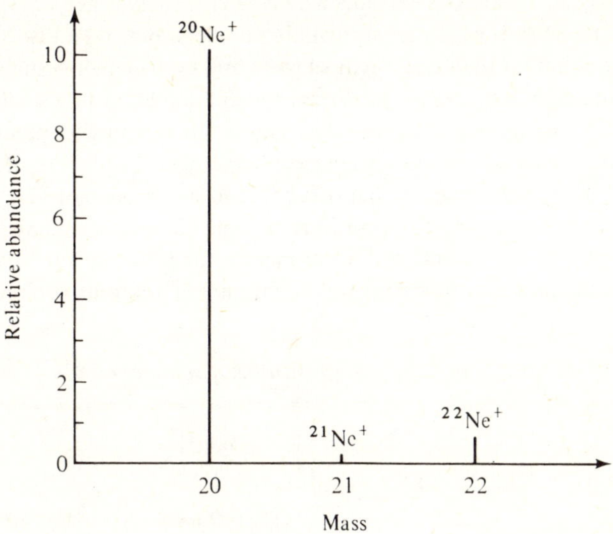

6.3 Mass spectrum of neon

These are called isotopes of neon; it is now known that atoms of all three neon isotopes contain 10 protons, but that they have different numbers of neutrons: 10, 11 and 12 respectively. The chemistry of an element is determined by the number of protons it contains, and so the isotopes of neon are chemically indistinguishable. Remember in interpreting mass spectra that each peak corresponds to an ion of charge +1.

From the mass spectra of elements we can measure not only the masses of their isotopes, but also their abundances. The results for Ne are:

Mass	*Abundance* (%)
20	90.92
21	0.26
22	8.82

From these figures we can calculate the average relative atomic mass (r.a.m.)

of neon: the mass of each isotope is multiplied by its abundance, and the figures for all the isotopes are added together. Thus for Ne:

$$\text{Average r.a.m.} = \left(20 \times \frac{90.92}{100}\right) + \left(21 \times \frac{0.26}{100}\right) + \left(22 \times \frac{8.82}{100}\right)$$
$$= 18.18 + 0.06 + 1.94$$
$$= \underline{20.18}$$

The average relative atomic mass is the value which would be obtained from any determination on a bulk sample of an element, such as a measurement of the vapour density.

The existence of isotopes explains why elements such as neon have relative atomic masses which are not whole numbers. But in fact isotopes are rather more common than this might suggest, and even elements such as hydrogen and carbon have isotopic forms which are present to a small extent in all naturally occurring samples. This is one reason why precise relative atomic masses of elements are never exact whole numbers, although we shall see later in the chapter that other factors are important here too. Table 6.1 gives the isotopic compositions of some common elements. Note that bromine does not consist of ^{80}Br atoms, as might be expected from its relative atomic mass, but of a mixture of ^{79}Br and ^{81}Br atoms in almost equal proportions.

Table 6.1 *Isotopic compositions of some common elements*

Hydrogen	^{1}H	99.99%	^{2}H	0.01%		
Carbon	^{12}C	98.9%	^{13}C	1.1%		
Oxygen	^{16}O	99.76%	^{17}O	0.04%	^{18}O	0.20%
Chlorine	^{35}Cl	75.5%	^{37}Cl	24.5%		
Bromine	^{79}Br	50.5%	^{81}Br	49.5%		

For the great majority of elements the isotopic composition seems to vary very little from one sample to another, and so the average relative atomic masses of these elements are constant. Evidence from meteors and the light emitted by stars suggests that this may be true not only on Earth but across the Galaxy, and it is quite possible that for many elements isotopic compositions are roughly constant throughout the Universe.

Elements which are involved in radioactive decay chains do not always show constant isotopic composition however: the most important example of this is lead. The relative atomic mass of lead was for some years a source of dispute among scientists. Different workers obtained values which differed from each other by more than one atomic mass unit, and it eventually became clear that samples of lead from different areas simply had different relative atomic masses. We now know that lead which is found in rocks which are non-radioactive has a relative atomic mass of 207.2, with the most common isotopes being ^{206}Pb (25%), ^{207}Pb (22%) and ^{208}Pb (52%). But, when lead is obtained from rocks which also contain uranium, the relative atomic mass is lower because the proportion of ^{206}Pb is higher. Lead from thorium-containing rocks, on the other hand, has a higher relative atomic mass as it contains a higher proportion of ^{208}Pb. A study of the radioactive decay chains soon shows the reason for this variation. The commonest isotope of uranium is ^{238}U, and this decays over millions of years by a series of steps to lead, giving the ^{206}Pb isotope only; the average relative atomic mass of the lead present in uranium-containing rocks therefore decreases as the uranium decays. Similarly thorium exists mostly as ^{232}Th, and this slowly decays giving ^{208}Pb only; lead present in thorium-containing rocks therefore has an unusually high relative atomic mass.

Applications. Mass spectrometry has been used very widely over the years, and is now employed routinely in the identification of unknown substances. But here we shall concentrate on two particular applications in which mass spectrometry gives important information which is not readily available from any other methods.

Accurate measurements of isotopic composition have been used to help determine the age of rock samples. Three main methods have been used involving the systems:

(a) uranium–thorium–lead
(b) potassium–argon
(c) rubidium–strontium

We shall consider only the third system, which dates rocks which contain rubidium, but the principle involved is the same in all three cases and is also similar to that used in radiocarbon dating. Naturally occurring rubidium contains two isotopes, ^{85}Rb (72%) and ^{87}Rb (28%). The former is stable indefinitely, but ^{87}Rb undergoes slow radioactive decay to ^{87}Sr. The half-life for this decay is 5×10^{10} years, which means that every 5×10^{10} years the number of ^{87}Rb atoms in a sample decreases by a factor of two. The changes in concentration of ^{87}Rb and ^{87}Sr atoms with time are shown in figure 6.4. As we shall see in the next section, the masses of ^{87}Rb and ^{87}Sr are not exactly equal to each other, and accurate mass spectrometry provides a ready method of determining the relative amounts of each isotope. After a rock has been formed, there is no possibility of it gaining or losing atoms, and so, provided that the rock contained little or no strontium originally, measurement of the ^{87}Rb:^{87}Sr ratio allows the age of the rock to be found from figure 6.4.

If the rock originally contained strontium as well as rubidium, then the situation is a little more complicated, as naturally occurring strontium contains 7.02% ^{87}Sr. A correction now has to be made for the ^{87}Sr which has not been

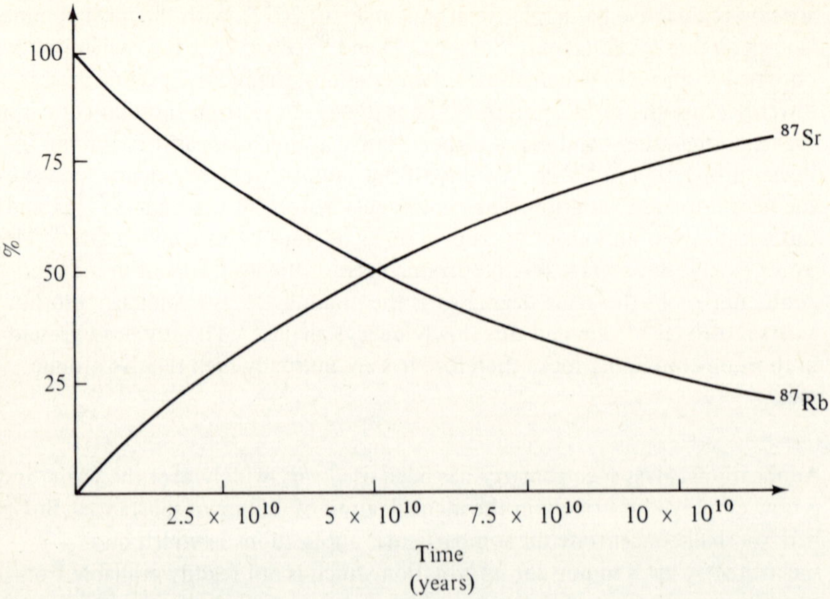

6.4 *Radioactive decay of ^{87}Rb to ^{87}Sr*

formed by radioactive decay, and once again mass spectrometry is used. The ratio of $^{86}Sr : ^{87}Sr$ in non-radioactive samples is measured, and the amount of ^{87}Sr originally present in the rock is then found by measuring the concentration of ^{86}Sr; here the important point is that the decay of rubidium produces ^{87}Sr, but no ^{86}Sr.

A different application of mass spectrometry lies in the field of organic reaction mechanisms. If we consider the hydrolysis of an ester such as ethyl ethanoate:

$$CH_3COOC_2H_5 + H_2O = CH_3COOH + C_2H_5OH$$

then clearly a C–O bond must be broken during the reaction, but there are two possibilities:

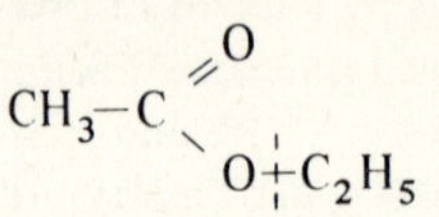

or

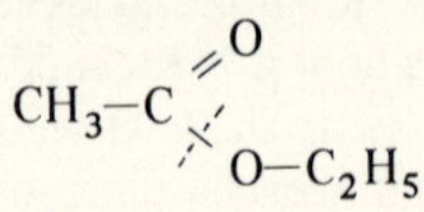

Feasible reaction mechanisms could be written for both of these alternatives; we can discover which actually occurs by the technique of isotope labelling. Ethyl ethanoate is prepared with one of its oxygen atoms consisting of the ^{18}O isotope:

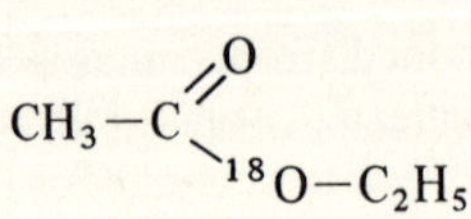

Now if it is the $CH_3CO\text{-}OC_2H_5$ bond which breaks during the hydrolysis, then the reaction will produce:

$$CH_3{-}C\begin{matrix}{\nearrow\!\!\!\!\nearrow} O\\ {\searrow} {}^{18}O{-}C_2H_5\end{matrix} \quad + \; H_2O \quad = \quad CH_3{-}C\begin{matrix}{\nearrow\!\!\!\!\nearrow} O\\ {\searrow} OH\end{matrix} \quad + \; C_2H_5^{18}OH \qquad (1)$$

But if the $CH_3COO\text{-}C_2H_5$ bond breaks, then:

$$CH_3{-}C\begin{matrix}{\nearrow\!\!\!\!\nearrow} O\\ {\searrow} {}^{18}O{-}C_2H_5\end{matrix} \quad + \; H_2O \quad = \quad CH_3{-}C\begin{matrix}{\nearrow\!\!\!\!\nearrow} O\\ {\searrow} {}^{18}OH\end{matrix} \quad + \; C_2H_5OH \qquad (2)$$

The reaction mixture is now analysed in a mass spectrometer: the relative molecular mass of the acid is found to be 60, which indicates that it is reaction (1) which occurs; reaction (2) would lead to the acid having relative molecular mass 62.

Isotopic clusters. The existence of isotopes can be of considerable assistance in interpreting the mass spectra of unknown compounds. The elements chlorine and bromine give rise to features in mass spectra which are easily recognised, as they both have two significant isotopes.

Figure 6.5 shows part of the mass spectrum of CH_3Br; peaks with low intensity have been omitted. The molecule of $CH_3{}^{79}Br$ has mass 94, and $CH_3{}^{81}Br$ has mass 96; the spectrum therefore contains two peaks separated by two mass units. The peaks have equal intensities as the abundances of ^{79}Br and ^{81}Br are almost equal; this pattern is characteristic of all compounds which contain one bromine atom.

A different pattern is obtained for molecules which contain two bromine atoms. If we consider CH_2Br_2, there are now three isotopic molecules,

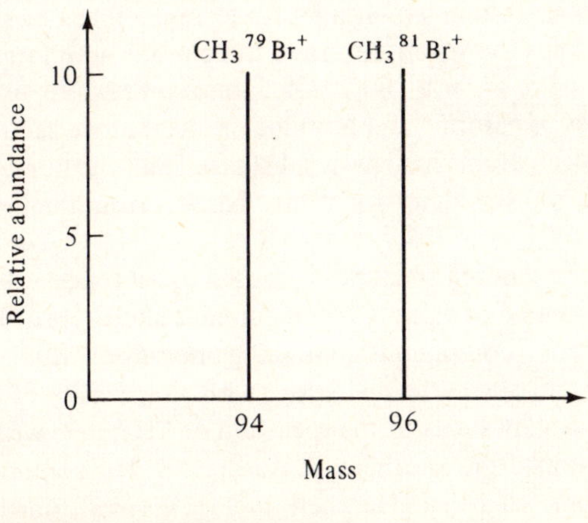

6.5 *Part of mass spectrum of CH_3Br*

$CH_2{}^{79}Br_2$, $CH_2{}^{79}Br^{81}Br$ and $CH_2{}^{81}Br_2$, with masses 172, 174 and 176 respectively; the mass spectrum of CH_2Br_2 is shown in figure 6.6. The bromine isotopes are distributed at random throughout the molecules, and so $CH_2{}^{79}Br^{81}Br$ occurs twice as often as either $CH_2{}^{79}Br_2$ or $CH_2{}^{81}Br_2$; in exactly the same way, when a coin is tossed twice, the probability of heads appearing twice is only half that of heads and tails appearing once each. The 1 : 2 : 1 intensity pattern, with the peaks separated by two mass units, is characteristic of compounds with two bromine atoms. Compounds with three bromine atoms show four peaks, with intensities in the ratio 1 : 3 : 3 : 1, etc.

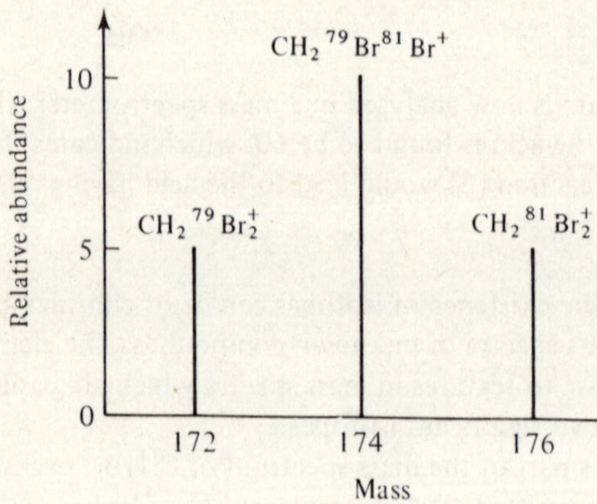

6.6 Part of mass spectrum of CH_2Br_2

Chlorine compounds have mass spectra which resemble those of bromine compounds, but the intensity distribution among the peaks is now different, as the ^{35}Cl isotope is three times as abundant as ^{37}Cl. Figure 6.7 shows the mass spectrum of HCl, which contains two peaks with intensities in the ratio 3:1. The intensity distribution becomes more complex when there is more than one chlorine atom in a molecule: for example it can be shown that with two chlorine atoms the spectrum shows three peaks with intensities in the ratio $1 \times 3^2 : 2 \times 3^1 : 1 \times 3^0$, or 9 : 6 : 1. The important point is to see that the existence of isotopes of chlorine and bromine means that we can readily identify the number of halogen atoms in a given substance from the number of peaks in the mass spectrum, and the identities of the halogens from the intensity distributions.

So far we have ignored small peaks caused by isotopes of low abundance, but even these can be of value. Carbon contains about 1% of the isotope ^{13}C: therefore in organic compounds containing one carbon atom, there is a small peak one mass unit above the main peak with an intensity of 1% of the main peak. Figure 6.8 shows part of the spectrum of CH_4; here we have neglected small contributions from another rare isotope ^2H. If a compound contains two carbon atoms, such as C_2H_6, then three peaks are formed: $^{12}C_2H_6$ has mass 30, $^{12}C^{13}CH_6$ has mass 31, and $^{13}C_2H_6$ mass 32. The intensities of

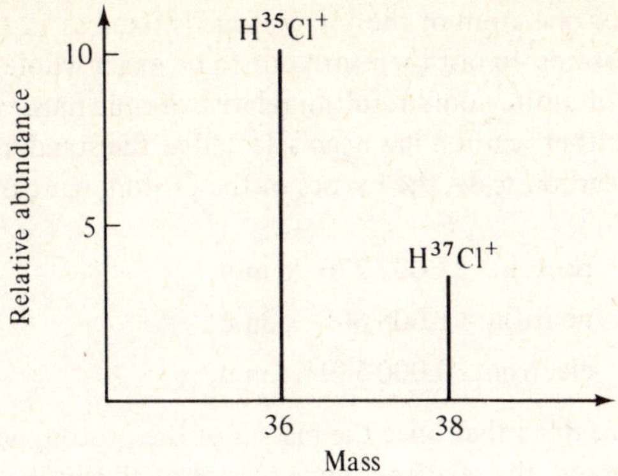

6.7 *Part of mass spectrum of HCl*

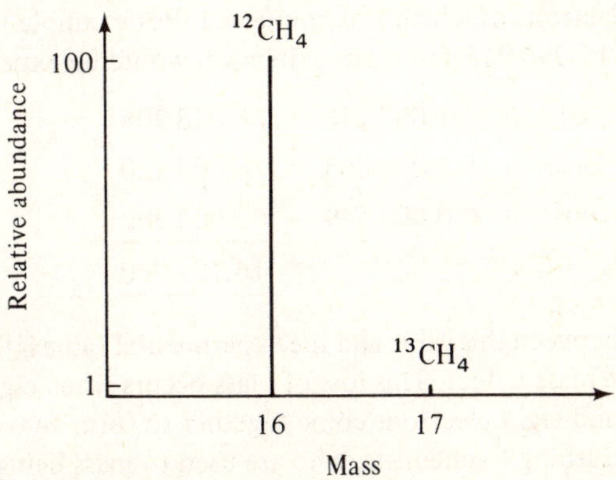

6.8 *Part of mass spectrum of CH_4*

these peaks are in the ratio 100 : 2 : 0.01, and in practice this means that the third peak is not detected. In general, if an organic compound contains *n* carbon atoms, then there is a peak one mass unit above the main peak with intensity approximately *n*% of that of the main peak. Measurement of the intensity of this peak provides a convenient method of identifying the number of carbon atoms in a compound.

C Accurate mass spectrometry

We have already seen that mass spectrometry is capable of measuring the masses of atoms and molecules with considerable accuracy; this section is concerned with the information provided by such measurements. The most obvious result is that the masses of individual atoms and molecules are not exact multiples of the mass of a hydrogen atom, and it therefore becomes necessary to define rather carefully what is meant by the scale of relative atomic masses. For accurate work it is normal to use the carbon scale, on which the mass of one atom of the ^{12}C isotope is fixed as 12.000 000. The masses of other atoms do not then turn out to be exact whole numbers, but no other choice of atom would result in relative atomic masses being exact whole numbers either: carbon has been selected as the standard for practical reasons. On the carbon scale, the masses of the proton, neutron and electron are:

proton:	1.007 276	a.m.u.
neutron:	1.008 665	a.m.u.
electron:	0.000 549	a.m.u.

It might be imagined that once the masses of the proton, neutron and electron were known, the relative atomic masses of all isotopes could be calculated simply by addition. But it is found that the mass of an atom is usually slightly less than would be predicted from the masses of the protons, neutrons and electrons of which it is composed. For example the mass of an atom of ^{16}O is 15.994 915 a.m.u.; now its mass would be expected to be:

8 protons:	8 × 1.007 276 =	8.058 208
8 neutrons:	8 × 1.008 665 =	8.069 320
8 electrons:	8 × 0.000 549 =	0.004 392
		16.131 920

The difference between this value and the experimental value is 0.137 005 a.m.u., and is called the mass defect. This loss of mass occurs when eight protons, eight neutrons and eight electrons come together to form an oxygen atom; this may be perturbing to chemists, who are used to mass being conserved in chemical reactions, but it is quite familiar to nuclear physicists, who see mass and energy as being interconvertible. The energy (E) evolved in a nuclear reaction, such as the formation of an oxygen atom, is related to the mass defect (m) by Einstein's equation:

$$E = mc^2 \qquad \text{(where } c \text{ is the speed of light)}$$

Now for the formation of oxygen, $m = 0.137$ a.m.u., or $0.137/(6 \times 10^{26})$ kg, and as $c = 3 \times 10^8 \text{ m s}^{-1}$:

$$E = \frac{0.137 \times (3 \times 10^8)^2}{6 \times 10^{26}} \text{ J atom}^{-1}$$

$$= \frac{0.137 \times (3 \times 10^8)^2 \times 6 \times 10^{23}}{6 \times 10^{26}} \text{ J mol}^{-1}$$

$$= 1.2 \times 10^{13} \text{ J mol}^{-1}$$

$$= \underline{1.2 \times 10^{10} \text{ kJ mol}^{-1}}$$

This is a huge amount of energy, far exceeding that produced in any chemical reaction. In general nuclear reactions almost always generate much more energy than chemical reactions, and this is of course reflected in the relative powers of nuclear and conventional weapons.

In fact the relationship between mass and energy applies to chemical as well as nuclear reactions, and so we can calculate the change in mass in a typical exothermic reaction. If we consider:

$$CH_4 + 2O_2 = CO_2 + 2H_2O$$

then roughly:

$$E = 900\,\text{kJ mol}^{-1}$$
$$= 9 \times 10^5\,\text{J mol}^{-1}$$

Therefore

$$m = \frac{9 \times 10^5}{(3 \times 10^8)^2}\,\text{kg mol}^{-1}$$
$$= 10^{-11}\,\text{kg mol}^{-1}$$
$$= 10^{-8}\,\text{g mol}^{-1}$$

As the mass of one mole of reactants is 80 g, the change in mass is about 10^{-8} %. It is therefore clear that for most practical purposes the law of conservation of mass is a good approximation for chemical reactions.

There are therefore three reasons why the relative atomic masses of elements are not exact whole numbers. Firstly the masses of the proton and the neutron are not exactly equal to one, nor is the mass of the electron exactly zero. Secondly the mass of an atom is not exactly equal to the masses of the protons, neutrons and electrons of which it is composed, because of the mass defect. Thirdly for many elements there is more than one stable isotope, and the relative atomic mass is therefore an average of the values of the stable isotopes.

Just as the heat of formation of a chemical compound gives us information on the stability of the compound, so the mass defect of an atom gives information on the stability of its nucleus: accurate measurements of relative atomic masses are therefore of great value to nuclear physicists. We cannot consider the stability of nuclei in any detail in this book, but we may note the variation in nuclear stability with atomic number, which is shown in figure 6.9. On the y-axis is plotted mass defect/atomic mass, which gives a measure of the average binding energy per nucleon. The graph shows that the most stable nuclei are those with intermediate atomic numbers; very heavy and very light nuclei have lower average binding energies. Energy is therefore evolved when a very heavy nucleus splits into two or more lighter nuclei: this is nuclear fission, and is the basis on which nuclear power stations operate. Energy is also evolved if two very light nuclei combine to form a heavier nucleus: this is nuclear fusion, and is the source of the energy of the sun.

The use of accurate mass spectrometry is not restricted to nuclear physicists however, and it is employed in analytical chemistry for the identification of unknown substances. The principle of the method can be

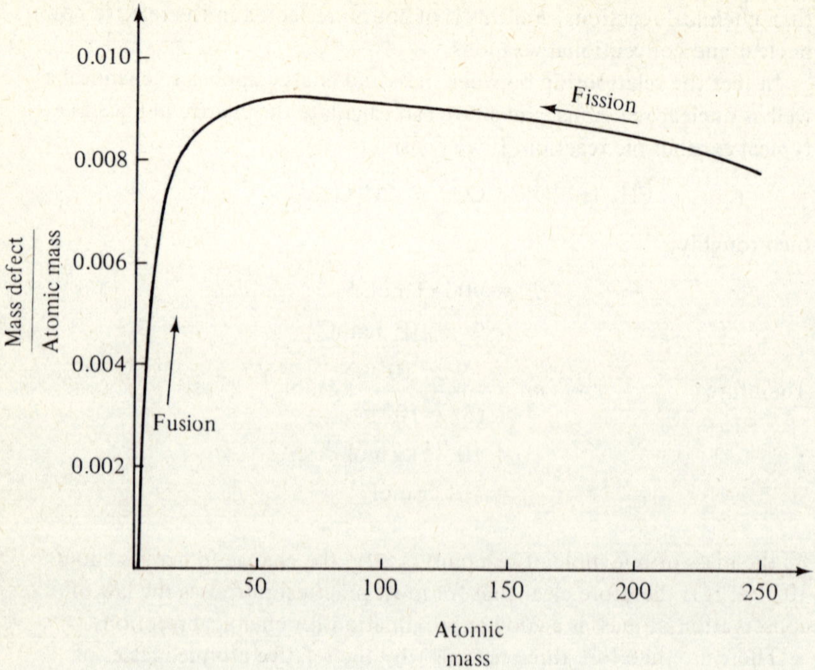

6.9 *Nuclear binding curve*

seen from a simple example. Suppose we are trying to identify a substance produced in an organic reaction. Mass spectrometry is an obvious technique to use, and if we find the relative molecular mass is 84, then the compound must be one of C_6H_{12}, C_5H_8O, $C_4H_4O_2$, $C_4H_8N_2$ and C_4H_4S. Of course simple chemical arguments may be of some help: for example the reaction mixture may not contain any sulphur, which would eliminate C_4H_4S. But only in the very simplest cases does determination of the relative molecular mass provide a unique answer, and usually the extra information provided by accurate mass spectra is required. The exact relative molecular masses of the five compounds above can be calculated from the relative atomic masses of the relevant isotopes: $^{1}H = 1.007\,825$, $^{12}C = 12.000\,000$, $^{14}N = 14.003\,074$, $^{16}O = 15.994\,915$ and $^{32}S = 31.972\,074$. Remember that there is no mass defect when a chemical compound is formed from its elements. So for C_6H_{12},

$$\begin{aligned}\text{r.m.m.} &= (6 \times 12.000\,000) + (12 \times 1.007\,825)\\ &= 84.093\,900\end{aligned}$$

Now if this calculation is performed for each of our five possible compounds, the results are:

C_6H_{12}	84.093 900
C_5H_8O	84.057 515

$C_4H_4O_2$	84.021 130
$C_4H_8N_2$	84.068 748
C_4H_4S	84.003 374

It is therefore clear that an accurate determination of the relative molecular mass will allow the definite identification of our unknown substance. Note, however, that we cannot yet distinguish between isomers such as cyclohexane and hex-1-ene, which both have the formula C_6H_{12}.

D Mass spectra of molecules

In mass spectrometers the energy of the bombarding electrons is usually chosen to be about 7000 kJ mol^{-1}. This is much greater than a typical molecular ionisation energy, and so the bombarding electrons have enough energy to knock an electron off any molecule with which they collide. But during a collision a bombarding electron may also transfer enough energy to a target molecule to allow a covalent bond in the molecule to be broken subsequently; remember that covalent bond energies are only of the order of a few hundred kJ mol^{-1}. Thus when methane is placed in a mass spectrometer, the primary step is:

$$CH_4 + e^- = CH_4^+ + 2e^-$$

But the CH_4^+ ion may have received enough energy to break one of its C–H bonds, so that the following secondary step may occur:

$$CH_4^+ = CH_3^+ + H$$

Further bond fission can result in the formation of CH_2^+ and even CH^+ ions.

These secondary processes have an important effect on the appearance of the mass spectrum of methane. Some CH_4^+ ions undergo no bond fission for a sufficiently long time to allow them to pass right through the spectrometer, and they give rise to a peak at mass 16. Others decompose before they enter the electric and magnetic fields; if they form CH_3^+ ions and H atoms, then they give a peak at mass 15. Alternatively, if they form H^+ ions and CH_3 radicals, then a peak at mass 1 appears. Of course the uncharged products H and CH_3 are not deflected by the magnetic field, and so do not give rise to peaks in the mass spectrum. A few ions undergo further decomposition into CH_2^+ ions, and they give a smaller peak at mass 14. Ions which decompose while in the electric or magnetic fields do not give rise to any distinct peak, and we may ignore them. Figure 6.10 shows the full mass spectrum of methane. Remember in interpreting this and other mass spectra that each peak corresponds to an ion of charge +1; do not try to use chemical intuition, or ideas about oxidation numbers, to predict the charges on each ion.

Fragmentation patterns, such as that shown for methane, are found in the spectra of all molecules, even though they have been carefully omitted from some of the spectra earlier in this chapter. At first this might seem to be just an irritating source of complexity; but in practice it is fairly easy to explain the appearances of the spectra of even quite large molecules, and furthermore the fragment peaks contain information which allows us to deduce details of the molecular structure of an unknown compound. Some simple examples will illustrate how the peaks of mass spectra are assigned, and then used to obtain structural information.

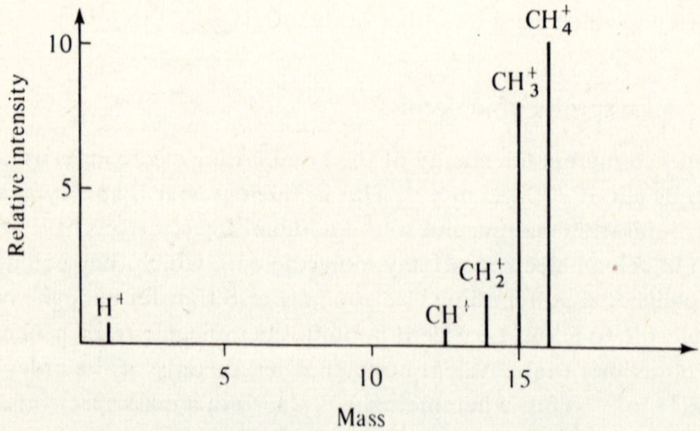

6.10 *Mass spectrum of CH_4*

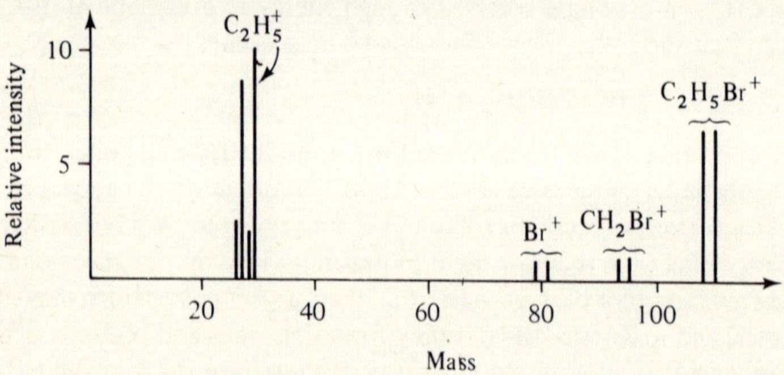

6.11 *Mass spectrum of C_2H_5Br*

Figure 6.11 shows the mass spectrum of C_2H_5Br. Bromine has two isotopes ^{79}Br and ^{81}Br, each with an abundance of 50%: electron bombardment therefore produces equal quantities of $C_2H_5\,^{79}Br^+$ and $C_2H_5\,^{81}Br^+$ ions, which have masses 108 and 110. All the other peaks in the spectrum arise from fragmentation of these ions, and therefore occur at lower mass numbers. In assigning these peaks, remember that the commonest form of fragmentation simply involves the breaking of a single covalent bond. If the C–C bond breaks, then we could have:

$$CH_3CH_2Br^+ \longrightarrow CH_3 + CH_2Br^+$$

This gives rise to two peaks, due to $CH_2{}^{79}Br^+$ (mass 93) and $CH_2{}^{81}Br^+$ (mass 95). Breaking the C–Br bond could lead to the loss of a C_2H_5 group:

$$CH_3CH_2Br^+ \longrightarrow CH_3CH_2 + Br^+$$

So we see peaks at 79 and 81 due to Br^+. Alternatively the products could be:

$$CH_3CH_2Br^+ \longrightarrow CH_3CH_2^+ + Br$$

This gives rise to the peak at mass 29. The other peaks with masses 26, 27 and 28 are due to the hydrocarbons $C_2H_2^+$, $C_2H_3^+$ and $C_2H_4^+$, which are formed by further fragmentation of $C_2H_5Br^+$.

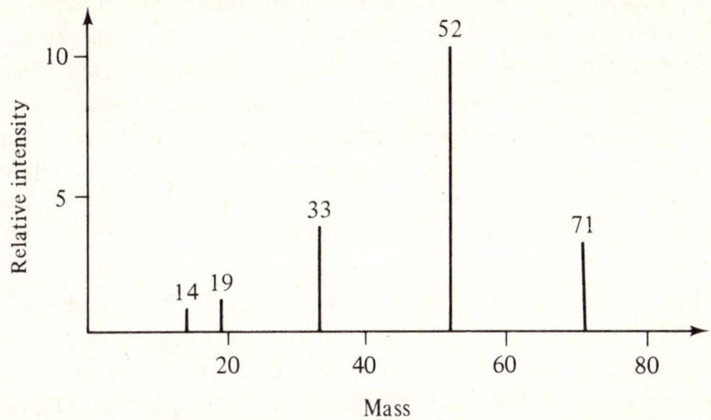

6.12 Unidentified mass spectrum (see text for assignments)

Figure 6.12 shows another mass spectrum; in this case we can deduce the identity of the compound. The highest peak is at mass 71, and the other peaks occur at masses 52, 33, 19 and 14. Peaks with very low intensities due to uncommon isotopes are not shown. We would expect the relative molecular mass of our compound to be 71; only in rare cases does the peak corresponding to the relative molecular mass have too low an intensity to be detected. In assigning the fragmentation peaks it is often useful to consider not just the mass of a peak, but the difference in mass between it and the highest peak. Thus the peak at 52 has a mass 19 less than the molecular peak at 71, and is therefore formed by the molecule losing an atom or group of atoms of mass 19. Now no simple chemical groups have mass 19, but that is the relative atomic mass of fluorine, so we may conclude that the peaks at 52 and 71 differ by one fluorine atom. The peak at mass 33 lies 38 below the molecular peak at 71, and no atoms or simple groups have this mass; however, as it lies 19 below the peak at 52, it could correspond to the original molecule minus two fluorine atoms. In exactly the same way the peak at 14 could correspond to the original molecule minus three fluorine atoms; from its mass we may also suggest that it is simply a nitrogen ion, so that our unknown compound is therefore NF_3. The remaining peak at 19 is due to F^+ ions. The full assignment is:

71	NF_3^+	52	NF_2^+
33	NF^+	19	F^+
14	N^+		

Notice that the fragmentation pattern has allowed us to identify the sample in this case; simply determining the relative molecular mass would not distinguish between NF_3 and C_4H_9N for example, unless we went to the considerable trouble (and expense) of a very accurate measurement.

The fragmentation patterns of organic molecules have been studied in great detail, and as a result it is possible to deduce information about the structural formulae of molecules just from their fragmentation patterns. We cannot go into this topic in detail, but an example will illustrate the power of the method. Suppose we are trying to discover whether a given substance is CH_3CH_2CHO or CH_3COCH_3. These two compounds are isomers, and therefore have exactly equal relative molecular masses. They both give peaks due to their molecular ions at 58, and we may predict that each could give a peak at 43 (= 58 − 15) corresponding to the molecular ion losing a methyl group of mass 15. CH_3CH_2CHO should also give a peak at 29(= 58 − 29) corresponding to the loss of an ethyl group of mass 29, but CH_3COCH_3 will not do this, as it has no ethyl group to lose. We should therefore be able to distinguish between our two isomers by looking for a peak at mass 29. Another simple test is to look for a peak at mass 57(= 58 − 1), corresponding to the loss of a hydrogen atom. Hydrogen atoms attached to C=O groups are very readily lost, whereas those in alkyl groups are more difficult to remove. CH_3CH_2CHO should therefore have a strong peak at 57, whereas CH_3COCH_3 should not. Figure 6.13 shows the actual spectra of CH_3CH_2CHO and CH_3COCH_3, which should be compared with our predictions. Notice that we have not bothered to assign all the lines in either spectrum: looking for a few key features has given the structural information required.

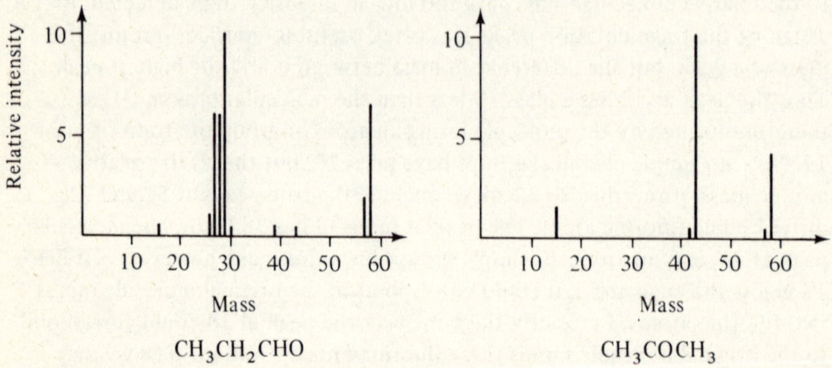

6.13 Mass spectra for propanal and propanone

7 Free radicals in the atmosphere

A NO and NO_2

One of the unusual features of the chemistry of nitrogen is the very large number of compounds which it forms with oxygen: at least five chemically distinct oxides have been characterised, and there may be others of rather low stability. The formulae of the main oxides are given in table 7.1, and we can see that all the oxidation states of nitrogen from +1 to +5 are represented.

Table 7.1 *The oxides of nitrogen*

Formula	*Common name*	*Oxidation state*	*Electrons*
N_2O	Nitrous oxide	+1	22
NO	Nitric oxide	+2	15
N_2O_3	Dinitrogen trioxide	+3	38
NO_2	Nitrogen dioxide	+4	23
N_2O_4	Dinitrogen tetroxide	+4	46
N_2O_5	Dinitrogen pentoxide	+5	54

We should not be surprised that nitrogen forms more than one oxide, for variable valency is quite common among the non-metallic elements. For example sulphur shows a variety of oxidation states in the compounds H_2S (−2), SO_2 (+4) and H_2SO_4 (+6), and chlorine forms a series of acids HCl (−1), HClO (+1), $HClO_2$ (+3), $HClO_3$ (+5), and $HClO_4$ (+7). What is unusual about the oxides of nitrogen is that two of the compounds, NO and NO_2, are odd-electron species. Although variable valency is common in other non-metals, stable free radicals are not, and so the properties of NO and NO_2 have few parallels in the chemistries of other elements. But NO and NO_2 are not only of interest because of their unusual electronic structures; they also play important roles in some of the chemical reactions which occur in the Earth's atmosphere, which we shall discuss later in this chapter.

No completely satisfactory explanation has ever been put forward for the existence of so many oxides of nitrogen; indeed, there is not even a simple method of representing adequately the chemical binding in either NO or NO_2. But we can get an important clue about the nature of the stability of these oxides from their heats of formation, which are shown in table 7.2. All the oxides except one are formed from the elements oxygen and nitrogen in endothermic reactions, and the reverse reactions, in which N_2 and O_2 are formed by the decomposition of the oxides, would therefore be energetically favourable. However the oxides are stable at room temperature because these decompositions occur at very slow rates; the oxides are said to be kinetically stable. It is important to remember here that there is no necessary connection between the enthalpy change for a chemical reaction and its rate. A similar explanation may be advanced for the stability of almost any organic compound in air: the reaction of sugar and air giving carbon dioxide and water is energetically favourable, as may be shown by adding sugar to an open fire, but of course the reaction does not proceed at any measurable rate at room temperature. The stabilities of the oxides of nitrogen may therefore be attributed to their inability to decompose into nitrogen and oxygen at any appreciable rate, rather than to any particular strength of their chemical bonds.

Table 7.2 *Heats of formation of the oxides of nitrogen*

Compound	*ΔH_f (298 K)*
N_2O	$+82\ kJ\ mol^{-1}$
NO	+90
N_2O_3	+84
NO_2	+33
N_2O_4	+9
N_2O_5	−43

Nitrogen dioxide, NO_2, may be readily prepared in the laboratory, either by the action of concentrated nitric acid on copper turnings:

$$Cu + 4HNO_3 = Cu(NO_3)_2 + 2NO_2 + 2H_2O$$

or by heating lead (II) nitrate:

$$2Pb(NO_3)_2 = 2PbO + 4NO_2 + O_2$$

It is a brown gas, and we can recognise its colour as being related to its odd number of electrons; other examples of coloured odd-electron species include the triphenylmethyl radical, $(C_6H_5)_3C\cdot$, solvated electrons, and transition metal ions. It also shows another characteristic property of odd-electron species, that is paramagnetism. At room temperature, NO_2 is in chemical equilibrium with its dimer N_2O_4:

$$2NO_2 \rightleftharpoons N_2O_4$$

In contrast to NO_2, N_2O_4 is not an odd-electron species, is colourless and is not paramagnetic. N_2O_4 is formed by the odd electrons in two NO_2 molecules forming a single covalent bond between the two nitrogen atoms. But this bond is very weak (57 kJ mol^{-1}), and can be broken during collisions even at room temperature.

Nitric oxide, NO, may also be prepared by the action of nitric acid on copper turnings; the use of 50% nitric acid yields mostly NO:

$$3Cu + 8HNO_3 = 3Cu(NO_3)_2 + 2NO + 4H_2O$$

The gas may also be prepared by adding hydrochloric acid to $NaNO_2$:

$$3NaNO_2 + 3HCl = 3NaCl + 2NO + HNO_3 + H_2O$$

Nitric oxide is colourless at room temperature, and not coloured as we might have expected, but it turns blue when liquefied or solidified. Its most familiar property is its rapid reaction with oxygen gas at room temperature, giving brown fumes of nitrogen dioxide:

$$2NO + O_2 = 2NO_2$$

It can therefore only be stored in the absence of air. Unlike NO_2, NO shows little tendency to dimerise although the solid may contain N_2O_2 molecules; the heat of dissociation of N_2O_2 to NO is only about 8 kJ mol^{-1}.

NO and NO_2 are clearly considerably more stable than most of the other free radicals which we have encountered in this book, such as $CH_3\cdot$ and $Cl\cdot$. Nevertheless they do show some reactions typical of the more reactive radicals. Thus NO reacts with the halogens forming nitrosyl halides:

$$2NO + Cl_2 = 2\,Cl{-}NO$$

This is analogous to the reaction of $CH_3\cdot$ with Cl_2. Similarly NO_2 is used as an oxidising agent in organic chemistry; the first step is hydrogen abstraction:

$$R{-}H + \cdot NO_2 = R\cdot + H{-}NO_2$$

This is similar to the reaction of $Cl\cdot$ with CH_4.

The dimerisation of NO_2. The dimerisation of NO_2 to N_2O_4 is a particularly good example for the study of chemical equilibria in the gas phase. In this section we shall be concerned with two problems: how the volume of a mixture of NO_2 and N_2O_4 varies with temperature, and how the proportion of NO_2 in a mixture of NO_2 and N_2O_4 varies with total pressure.

Figure 7.1 shows the variation of the volume of a sample of NO_2 gas with temperature; the pressure is kept constant at 1 atmosphere. The ideal gas laws predict that for a fixed mass of gas at constant pressure the volume is directly proportional to the temperature (in K), and so we should expect figure 7.1 to show a straight-line graph which goes through the origin. In

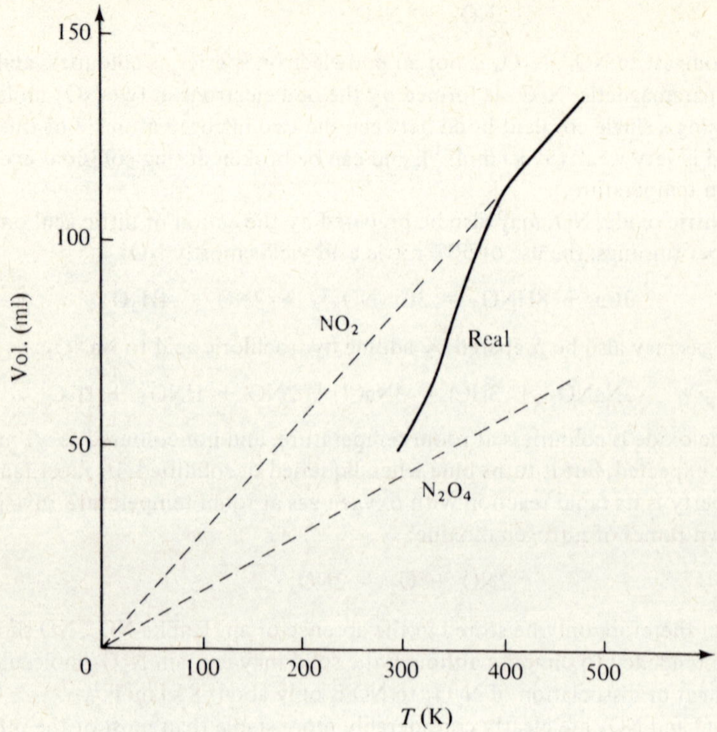

7.1 Variation of the volume of a sample of NO_2 with temperature

fact this is true only for the readings above about 410 K; at lower temperatures the volume is lower than we would expect from the ideal gas laws. The interpretation of the graph is not difficult: at temperatures above 410 K, the gas consists almost entirely of NO_2 molecules. At lower temperatures, there are appreciable amounts of N_2O_4 present in the equilibrium mixture, and the volume is therefore lower than expected as two molecules of NO_2 combine to form one molecule of N_2O_4. The lower the temperature, the greater is the deviation from the straight-line graph, and the greater is the proportion of N_2O_4. The line does not extend below 294 K as the gas liquefies at that temperature.

We can extract further information from the detailed shape of this graph. When the temperature is 330 K the volume of the gas is 70.5 ml. We can also see from the graph that, at that temperature, the volume would be 94 ml if no dimerisation to N_2O_4 occurred, and would therefore be $(\frac{1}{2} \times 94) = 47$ ml if the gas were fully dimerised. As 70.5 is halfway between 47 and 94, we may conclude that at 330 K, exactly half the molecules are NO_2, and half are N_2O_4. As the total pressure of the gas is 1 atmosphere, the partial pressure of each gas will be 0.5 atmosphere, and so we can calculate the value of the equilibrium constant:

$$K_p = \frac{P^2_{NO_2}}{P_{N_2O_4}}$$

$$= \frac{(0.5)^2}{0.5} \quad \frac{atm^2}{atm}$$

$$= 0.5 \quad atm$$

If we repeat this analysis for a series of points on the graph we can find how the equilibrium constant, K_p, varies with *T*. If we now plot a graph of $\log_e K_p$ against 1/T the result is a straight-line graph, as shown in figure 7.2. The

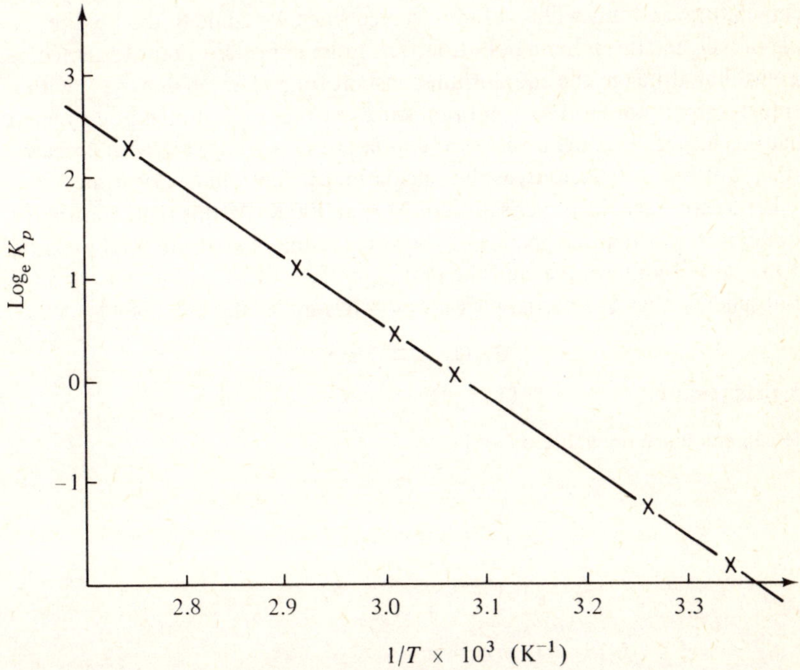

7.2 *Determination of* ΔH *for* $N_2O_4 \rightleftharpoons 2NO_2$

gradient of this graph is $-(\Delta H/R)$, where ΔH is the enthalpy change for the reaction $N_2O_4 = 2NO_2$, and *R* is the gas constant. Using the values from figure 7.2:

$$-\frac{\Delta H}{R} = -\frac{4.6}{0.67 \times 10^{-3}}$$

Therefore $$\Delta H = \frac{4.6 \times 8.31 \times 10^3}{0.67} \text{ J mol}^{-1}$$

$$= \underline{57 \text{ kJ mol}^{-1}}$$

If experiments are conducted on NO_2 gas above about 500 K it is found that further deviations from linearity occur, and the volume rises more rapidly than expected. Further dissociation of the NO_2 is now occurring, producing another increase in the number of molecules in the system:

$$2NO_2 = 2NO + O_2$$

The data may be treated just as above to give a value for the enthalpy change for this reaction too; it is found that this dissociation is not complete below about 1300 K.

So far we have considered the dimerisation of NO_2 at atmospheric pressure only. But the degree of dimerisation is affected by pressure as well as by temperature, and this will be of importance when we come to discuss the role of NO_2 in atmospheric pollution. The most important point to remember here is that although the equilibrium constant for a reaction does vary with temperature, it does not vary with pressure. At first sight it might be thought that this implies that the position of equilibrium does not vary with pressure either, but that is fallacious, as the calculations below will demonstrate.

Let us consider the dimerisation of NO_2 at 300 K. At this temperature the equilibrium constant has a value of about 0.17 atm. Now if the total pressure of the gas is 1 atmosphere, and the partial pressure of NO_2 molecules is x atmospheres, then the partial pressure of N_2O_4 molecules must be $(1 - x)$.

	$N_2O_4 \rightleftharpoons$	$2NO_2$	
Partial pressure:	$(1 - x)$	x	atm

We can calculate the value of x as:

$$K_p = \frac{x^2}{1 - x}$$

$$0.17 = \frac{x^2}{1 - x}$$

$$0.17 - 0.17x = x^2$$

Therefore $$0 = x^2 + 0.17x - 0.17$$

We therefore need to solve this quadratic equation, which may be irritating, but is not difficult:

$$x = \frac{-0.17 \pm \sqrt{(0.17^2 + 4 \times 0.17)}}{2}$$

$$= 0.34 \text{ or } -0.51$$

Rejecting the negative value as meaningless, we find that 34% of the molecules are NO_2 at equilibrium, and 66% are N_2O_4; therefore at atmospheric pressure there is substantially more N_2O_4 than NO_2 present.

But the position is different if we consider what happens if the pressure is reduced to 10^{-6} atmospheres. This is not as stupid as it may seem; we could do this in practice by taking some NO_2 at atmospheric pressure and increasing its volume by a million, but equally we could do it by injecting one part of NO_2 into 10^6 parts of some gas with which it did not react, such as air. In practice the atmosphere often contains about 1 in 10^6 parts NO_2, so this is a relevant question. If the partial pressure of NO_2 is again x atm, then the partial pressure of N_2O_4 will be $(10^{-6} - x)$ atm, and:

$$K_p = \frac{x^2}{(10^{-6} - x)}$$

Remembering that K_p does not change with pressure, we have:

$$0.17 = \frac{x^2}{10^{-6} - x} \qquad (1)$$

Therefore

$$0.17 \times 10^{-6} - 0.17x = x^2$$

This equation is best solved by approximation: x^2 is a very small term compared with the others, and neglecting it gives:

$$0.17 \times 10^{-6} - 0.17x = 0$$

Therefore $\underline{x = 10^{-6}}$

From **(1)** $$(10^{-6} - x) = \frac{x^2}{0.17} = \frac{10^{-12}}{0.17} = \underline{5.9 \times 10^{-12}}$$

This result now shows that when the concentration of NO_2 is very low, there is essentially no N_2O_4 present at all. The position of equilibrium has therefore changed, even though the equilibrium constant has remained the same. This is always the case, unless the reaction involves no change in the total number of gaseous molecules (such as $H_2 + I_2 = 2HI$). It is now clear that when we come to consider the role of NO_2 in the atmosphere, we need not be concerned at all with the possibility of NO_2 being dimerised, even though N_2O_4 is the main constituent of the gas at atmospheric pressure.

The reaction of NO with oxygen. It is well known that when nitric oxide is mixed with air, brown fumes of nitrogen dioxide are formed almost instantaneously:

$$2NO + O_2 = 2NO_2$$

The kinetics of this reaction have been studied in some detail, and it has been found that the reaction is second-order in NO, that is the rate of the reaction is proportional to the square of the NO concentration. To appreciate the significance of this result, we shall need to know a little about the half-lives of first and second-order reactions.

In first-order reactions, the rate is proportional to the concentration of one of the reactants. This results in the concentration of that substance showing the well-known exponential decay, which is illustrated in figure 7.3. The characteristic feature of exponential decay is that the half-life of the substance is constant: the time taken for the concentration to drop from 1 M to 0.5 M equals the time taken for the concentration to drop from 0.5 M to 0.25 M, and so on.

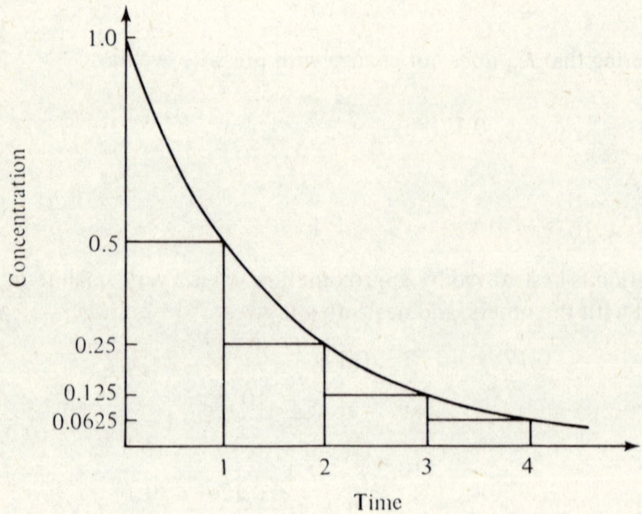

7.3 *First-order reaction kinetics*

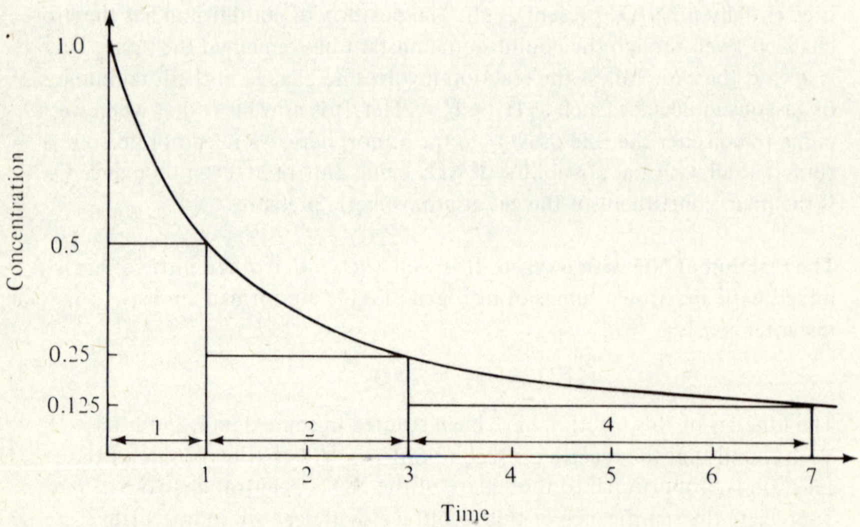

7.4 *Second-order reaction kinetics*

The decay curve for a second-order reaction is shown in figure 7.4. It appears qualitatively similar to that for a first-order reaction, but notice that the half-life is no longer constant. Instead the time taken for the concentration to go from 1 M to 0.5 M is only half that for it to go from 0.5 M to 0.25 M; the drop from 0.25 M to 0.125 M takes twice as long again. For second-order reactions the half-life is inversely proportional to the initial concentration.

Now the half-life for NO when a test-tube of the gas is mixed with air is about 0.1 s, which means that the NO is consumed very rapidly indeed. But if we take NO at a concentration of 10^{-6} atmospheres and mix it with air then as the half-life is inversely proportional to the initial concentration, it will be $0.1/10^{-6}$ s, or about a day. The reaction has now become very slow, and we shall see that when we consider the fate of NO molecules present in very small concentrations in the atmosphere, reaction with oxygen is not normally a significant factor.

B Photochemical smog

Over the last ten years there has been considerable interest in some of the chemical reactions which occur in the atmosphere. Even today our understanding of many of the processes is far from complete, but it is clear that many reactions involve free radicals, and that NO and NO_2 have an important influence, even though they are present in concentrations of less than one part per million.

One problem which has attracted much attention is smog. The characteristic features of smog are a reduction in visibility, irritation of the eyes and throat in humans, damage to and reduction in the growth of vegetation and, in severe cases, an increase in the death-rate. Although the phenomenon has been observed at many places in the world, Los Angeles has rightly become famous as the best point on Earth for the formation of smogs.

There are a number of factors which contribute to Los Angeles' undesired pre-eminence. The city's roads carry very heavy motor traffic, which emits small but significant amounts of hydrocarbons, NO and NO_2 into the atmosphere. The climate is very hot, and the city enjoys long periods of sunshine. Lastly the geography of the area is such that air tends to become trapped in a pocket over the city, dispersing only slowly. These factors combine to produce the classic recipe for smog – oxides of nitrogen, hydrocarbons and ultra-violet radiation.

The role of the oxides of nitrogen is quite well understood. NO_2, which is formed in motor vehicle exhausts, absorbs ultra-violet light of wavelength less than 380 nm; this causes an N–O bond to break, forming NO and an oxygen atom:

$$NO_2 + \text{u.v. light} = NO + O \qquad (1)$$

(Remember that we saw in the previous section that there is no question of NO_2 being dimerised to N_2O_4 at the low concentrations found in the atmosphere.) The oxygen atoms can react with O_2 molecules, which are of course present in high concentration, forming ozone:

$$O + O_2 = O_3 \qquad (2)$$

The ozone is then destroyed by NO molecules, either from equation (1) or directly from car exhausts:

$$NO + O_3 = NO_2 + O_2 \qquad (3)$$

(Remember that we saw in the last section that the reaction of NO with O_2 is not important when the concentration of NO is very low.) In the absence of other compounds the series of reactions (1) to (3) simply gives rise to a small steady concentration of O_3 and NO, and the NO_2 which is destroyed in reaction (1) is regenerated in reaction (3). But if hydrocarbons are present in the atmosphere, then this equilibrium between NO, NO_2 and O_3 is disturbed. Oxygen atoms and ozone molecules can react with alkenes producing organic free radicals such as $CH_3\cdot$ and $CH_3CO\cdot$, which undergo chain reactions which convert NO back to NO_2:

$$CH_3\cdot + O_2 = CH_3O_2\cdot$$
$$CH_3O_2\cdot + NO = CH_3O\cdot + NO_2 \qquad \text{(NO consumed)}$$
$$CH_3O\cdot + O_2 = CH_3O_3\cdot$$
$$CH_3O_3\cdot + NO = CH_3O_2\cdot + NO_2 \qquad \text{(NO consumed; } CH_3O_2\cdot \text{ regenerated)}$$

These chain reactions mean that equation (3) is no longer required to convert NO back to NO_2, and so the ozone concentration begins to rise.

A second result of the presence of hydrocarbons is the formation of peroxyacyl nitrates. $CH_3CO\cdot$ radicals, formed above, can react with O_2 molecules:

$$CH_3CO\cdot + O_2 = CH_3CO{-}O_2\cdot$$

Then with NO_2 molecules:

$$CH_3CO{-}O_2\cdot + NO_2\cdot = CH_3CO{-}NO_2$$

This terminates the chain reaction, as two free radicals combine to form a molecule with an even number of electrons, but the product $CH_3COO_2NO_2$, often called PAN, causes strong irritation to the eyes and throat even in tiny concentrations. A further result of the formation of organic free radicals is that polymerisation reactions, similar to those we saw in chapter 3, are initiated:

$$CH_3\cdot + CH_2{=}CH_2 = CH_3{-}CH_2{-}CH_2\cdot$$
$$CH_3{-}CH_2{-}CH_2\cdot + CH_2{=}CH_2 = CH_3{-}CH_2{-}CH_2{-}CH_2{-}CH_2\cdot \text{ etc}$$

The resulting polymers form tiny particles in the air and, as these scatter light efficiently, they produce a reduction in visibility.

Many other reactions can also occur in smogs, but it seems likely that the formation of ozone, PAN and tiny solid particles can explain most of the effects of smog on humans and plants. It is therefore clear that the best way to reduce the incidence of smog is to control the amounts of nitrogen oxides and hydrocarbons in the atmosphere. Efforts have therefore been made in Los Angeles to persuade people to use more public transport, thus decreasing the total amount of traffic on the roads, and much research has been directed to reducing pollutant emission from motor vehicles. The first of these has met with rather modest success, and it seems likely that control of exhaust gases may well be the only practical way of reducing smog.

The chemistry behind the formation of smog is now reasonably well understood, and considerable progress has already been made in reducing the frequency and severity of smogs. Some problems still remain: for example it is not entirely clear whether it is more important to reduce the emission of oxides of nitrogen or hydrocarbons from car exhausts. But it may be that no amount of legislation on car exhaust emission will ever eliminate smog entirely; oxides of nitrogen are produced by bacterial decomposition of organic matter, as well as by cars, and the major source of hydrocarbons in the atmosphere is not petrol but vegetation, which gives off unsaturated hydrocarbons called terpenes, and alkanes. As the only other ingredient required for photochemical smog is bright sunlight, it may well be that in future smogs will simply be natural rather than man-made.

C The ozone layer

Another topic which has received more than its fair share of attention in recent years is the ozone layer. The Earth is surrounded by a layer of ozone gas, which is of considerable importance to living matter as the ozone acts as a shield which protects the Earth's surface from damaging ultra-violet radiation. The ozone layer lies in the stratosphere, between 15 and 50 km above the Earth's surface, and contains up to 10 ppm ozone.

Ozone is formed in the stratosphere in the following way: O_2 molecules are split by ultra-violet light of very short wavelengths (< 242 nm) into oxygen atoms:

$$O_2 + \text{u.v. light} = O + O \tag{4}$$

These oxygen atoms can then combine with another O_2 molecule to form ozone:

$$O + O_2 = O_3 \tag{5}$$

Ozone can be destroyed by collision with an oxygen atom:

$$O_3 + O = O_2 + O_2 \tag{6}$$

It is also destroyed when it absorbs ultra-violet radiation of wavelengths less than 340 nm:

$$O_3 + \text{u.v. light} = O_2 + O \quad (7)$$

There is therefore a dynamic equilibrium in the stratosphere in which ozone is being continually formed and destroyed, and the relative rates of formation and destruction determine the ozone concentration.

Considerable attention has been paid to the question of whether the presence of any man-made pollutants in the stratosphere might alter the equilibrium concentration of ozone, thereby increasing the amount of ultra-violet radiation reaching the Earth's surface. This first reached the headlines in 1971 when the United States was deciding whether to build a supersonic aircraft, to rival the Anglo-French Concorde and the Russian TU-144. It was then suggested that supersonic aircraft, which would fly through the stratosphere, would deposit sufficient NO to cause a significant decrease in the amount of ozone, and thereby a significant increase in skin cancer on Earth. The suggestion was that the following reactions might occur:

$$NO + O_3 = NO_2 + O_2 \quad (8)$$

$$NO_2 + O = NO + O_2 \quad (9)$$

The overall result of these two reactions would be:

$$O_3 + O = O_2 + O_2 \quad (10)$$

which is the same as reaction (6). If reactions (8) and (9) take place significantly faster than equation (6), then the NO is acting as a catalyst for the reaction, as it is regenerated in reaction (9); that could produce a decrease in the amount of ozone in the stratosphere.

Unfortunately it was very difficult at the time to make an accurate estimate of the exact depletion of the ozone layer, but a well-publicised report suggested that the ozone layer might be halved by 1985. Partly as a result of fears of the environmental consequences, the United States decided not to go ahead with the production of supersonic aircraft. It is now generally agreed that the potential impact of supersonic aircraft was over-estimated. Technical improvements soon rendered estimates of the amount of NO produced by one aircraft unreasonably high, and further experiments and measurements on the stratosphere suggested that the ozone layer might be considerably more robust than some had imagined. As it happens the number of supersonic aircraft now in service is also much smaller than had been anticipated, and recent estimates suggest that the net effect of supersonic aircraft might be a slight increase in the ozone concentration.

Another popular scare about the ozone layer concerned, almost inconceivably, the use of aerosol cans. Fluorocarbons, for example CF_2Cl_2 and $CFCl_3$, found increasing use in the 1960s and 1970s as propellant gases in these. The main property which makes them suitable for that apparently harmless purpose is their great lack of reactivity, but ironically it was their very inertness which gave rise to fears about their safety. Because the fluorocarbons are so unreactive there is nowhere in the lower atmosphere where they are destroyed, and so they slowly diffuse upwards into the stratosphere. There they are irradiated with ultra-violet light, and dissociate:

$$CFCl_3 + \text{u.v. light} = CFCl_2\cdot + Cl\cdot$$

It is suggested that the Cl• atoms so formed might also catalyse the decomposition of ozone:

$$Cl\cdot + O_3 = ClO\cdot + O_2 \qquad (11)$$

$$O + ClO\cdot = Cl\cdot + O_2 \qquad (12)$$

$$O_3 + O = O_2 + O_2 \qquad (13)$$

Again the Cl• atoms are regenerated in reaction (12), and so reactions (11) and (12) act as propagation steps for a chain reaction for the destruction of ozone. As a result of this publicity the use of fluorocarbons is now restricted in some parts of the world, and fluorocarbon production, which rose steadily up to 1974, is now falling.

While many see a reduction in the use of fluorocarbons as a sensible precaution, and regard the prospect of the loss of the ozone shield as extremely serious, many scientists now believe that the dangers involved have been much exaggerated. We can identify three areas where uncertainties have arisen. Firstly our knowledge of exactly what reactions occur in the stratosphere, and their rates, is very patchy. Secondly it is very hard to measure accurately small changes in the ozone concentration. The amount of ozone in the stratosphere shows fluctuations of up to 20% throughout the year, and so trying to detect changes of less than one per cent per year is subject to considerable error. Thirdly the medical effects of a decrease in the ozone concentration are hard to estimate. Certainly some increases in skin cancer might occur, but whether this would be a more significant factor than changes due to increases in the popularity of sunbathing, or indeed than the Clean Air Acts, is less certain. Doubtless there will be further discussion on this vexed topic in years to come, but at present some cynics believe that it is more prudent to wear a hat against the sun than to worry about the ozone layer.

D Oxygen

At first sight oxygen would seem to be an unlikely molecule to feature in a book on odd-electron species: it contains 16 electrons, and its structure is commonly written O=O. Each atom therefore shares two pairs of electrons, giving it a complete shell. The bond lengths and bond energies shown in table 7.3 seem to confirm this description: the double bond in O_2 is shorter and stronger than the single bond in H_2O_2.

Table 7.3 *Properties of O=O and O–O bonds*

	O=O	H–O–O–H
Bond length (nm)	0.121	0.149
Bond strength ($kJ\ mol^{-1}$)	497	146

There is however one property of oxygen which is not consistent with it having a simple double bond, and which obliges us to look for another description. Oxygen is paramagnetic, that is each molecule behaves as if it were a tiny magnet. This may be shown simply by pouring liquid oxygen over a permanent magnet; each molecule is attracted by the magnet, and the oxygen is seen to cling to the poles. Paramagnetism is not a property of closed-shell molecules, and liquids such as ethanol, water and liquid nitrogen are not attracted to permanent magnets.

Paramagnetism is a result of the magnetic properties of the electron. Electrons behave as tiny magnets because they spin on their own axes; any charged particle which spins on its axis has a resultant magnetic moment. In all closed-shell atoms and molecules this is of no significance because, when any orbital is doubly occupied and therefore full, the spin of one electron is exactly opposed to the spin of the other and their magnetic moments cancel. Paramagnetism occurs only when one or more orbital is singly occupied; the commonest examples are NO, NO_2 and ClO_2, which each have an odd number of electrons and one singly-occupied orbital. In the case of oxygen there are two singly-occupied orbitals, so the molecule has a magnetic moment even though it has an even number of electrons.

Molecular orbital theory. To see why oxygen has two unpaired electrons, we shall need to use molecular orbital theory. The essential idea in this theory is that we may treat molecules in much the same way as atoms. In atoms electrons occupy atomic orbitals; the electrons are distributed so as to fill the most stable orbitals, subject to there being no more than two electrons in any one orbital. Thus the configuration of a sodium atom is:

$$1s^2\ 2s^2\ 2p_x^2\ 2p_y^2\ 2p_z^2\ 3s^1$$

(If the idea of $2p_x$, $2p_y$ and $2p_z$ orbitals is unfamiliar, see p. 74.) In molecules the position is the same: there are molecular orbitals, now spread over the whole molecule, and these are filled in order of their stability, with a maximum of two electrons per orbital.

Molecular orbitals are derived from the atomic orbitals of the atoms which make up a molecule; whenever two orbitals from different atoms overlap, they form two new molecular orbitals, one of which is more stable than the atomic orbitals, and one of which is less stable. The formation of a hydrogen molecule from two hydrogen atoms illustrates this principle.

When the atoms are widely separated each contains one electron in a 1*s* orbital. When the atoms approach the 1*s* orbitals overlap, and two new orbitals are formed, as shown in figure 7.5. There are two electrons in the hydrogen molecule, and so they both occupy the more stable orbital, which is labelled 1σ. They now have a lower energy than when the hydrogen atoms were separated, which explains why the H_2 molecule is stable. The 1σ orbital is called a bonding orbital, and the 2σ orbital an anti-bonding orbital.

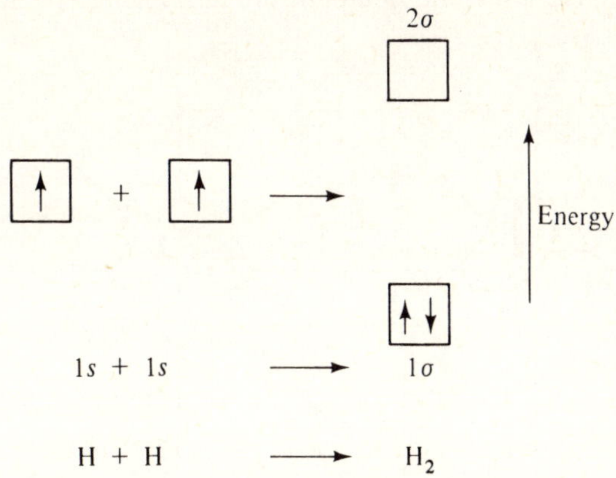

7.5 The formation of H_2 from H atoms

The case of the He_2 molecule is very similiar to that of H_2; figure 7.6 shows its molecular orbital diagram. The atomic orbitals of the He atoms are on the left and right sides of the diagram, and the resulting molecular orbitals in the centre. The orbitals involved are just the same as for the H_2 molecule, but there are four electrons in the He_2 molecule, so two occupy the 1σ molecular orbital, and two the 2σ. Two of the electrons are now more stable than they were in the separated atoms, but the other two are less stable, and

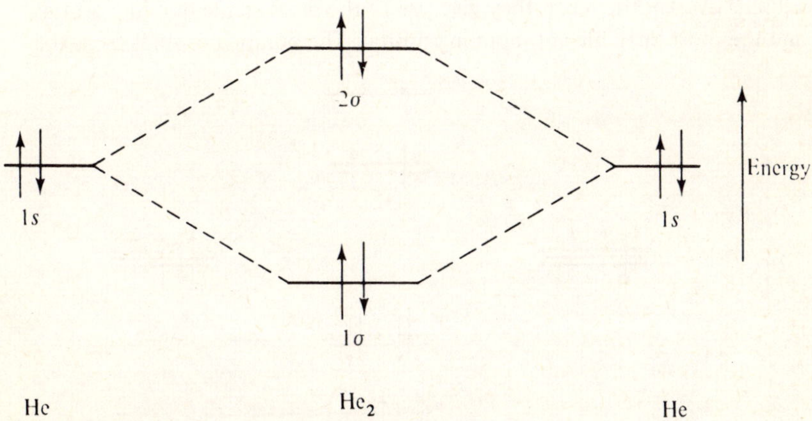

7.6 The formation of He_2 from He atoms

the net result is no change in energy when the He atoms are brought together. Put another way, there is no binding energy between the two atoms, which explains why the He_2 molecule is unknown, and He is a monatomic gas.

We can now go on to the bonding in the molecules N_2, O_2 and F_2. Just as the molecular orbital diagrams for H_2 and He_2 resembled each other, so those for N_2, O_2 and F_2 are qualitatively similar. We can neglect the 1*s* and 2*s* orbitals altogether, as they are all fully occupied in both the atoms and the molecules, and consider just the 2*p* atomic orbitals. Figure 7.7 shows the 2*p*

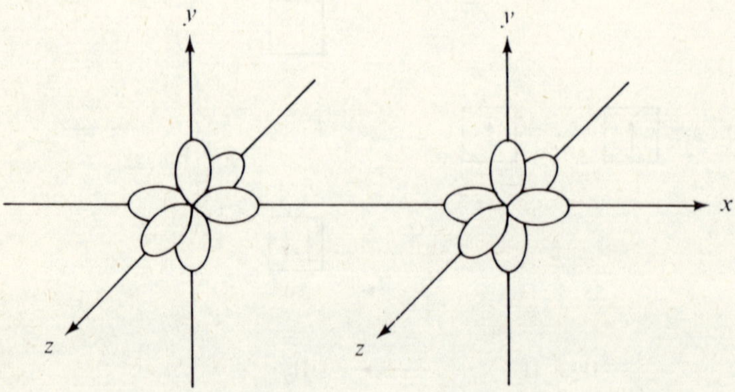

7.7 2p *orbitals on adjacent oxygen atoms*

orbitals on two adjacent oxygen atoms; the $2p_x$ orbitals point along the line between the two atoms, whereas the $2p_y$ and $2p_z$ orbitals are at right-angles to that line and each other. From the diagram it is clear that overlap between the $2p_x$ orbitals is end-on, and will be greater than that between the $2p_y$ and $2p_z$ orbitals, which is side-on. (Remember that orbitals are diffuse clouds, and do not have sharp edges.) Bearing in mind the points about overlap, we can now draw the molecular orbital diagram of O_2 (figure 7.8). As the $2p_x$ orbitals overlap the most, they give rise to the most stable bonding orbital, and the most unstable anti-bonding orbital. The bonding orbitals from the

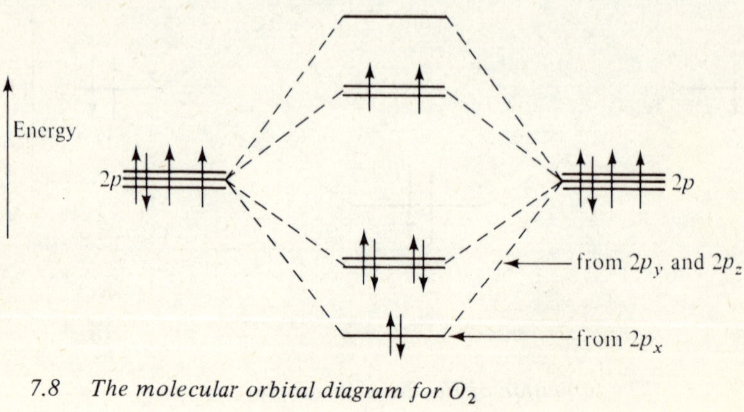

7.8 *The molecular orbital diagram for* O_2

$2p_y$ and $2p_z$ orbitals have the same energy as each other, by symmetry.

An oxygen molecule contains 16 electrons; 8 of these occupy the 1*s* and 2*s* orbitals on each atom, leaving 8 to occupy the orbitals in figure 7.8. Six of these fill the bonding orbitals from the 2*p* atomic orbitals, which leaves two electrons to occupy the anti-bonding orbitals. Now the two anti-bonding orbitals formed from the $2p_y$ and $2p_z$ orbitals have exactly the same energy, and so one electron occupies each orbital. (Double occupation of either orbital leads to greater electron–electron repulsion.) These two electrons may now have parallel spins, as they are not in the same orbital, therefore their magnetic moments do not cancel; and so O_2 is a paramagnetic molecule, with two unpaired electrons.

A comparison of the molecules N_2, O_2 and F_2 show that the presence of unpaired electrons is not incompatible with it having a double bond. Figure 7.9 shows the molecular orbital diagrams for each molecule; they differ only

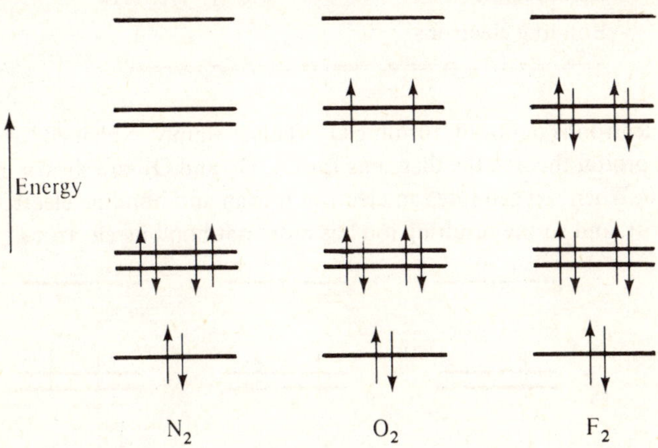

7.9 *The molecular orbital diagrams of* N_2, O_2 *and* F_2

in the number of electrons. Table 7.4 shows the number of electrons in bonding and anti-bonding orbitals in each molecule; it is clear that the net number of bonding electrons is consistent with the conventional description of the bonds as triple, double and single.

Table 7.4 *Molecular orbital description of* N_2, O_2 *and* F_2

Molecule	*Bonding electrons*	*Anti-bonding electrons*	*Difference*	*Bond*
N_2	6	0	6 – 0 = 6	Triple
O_2	6	2	6 – 2 = 4	Double
F_2	6	4	6 – 4 = 2	Single

The molecular orbital description of oxygen neatly explains another unusual feature of its chemistry. Covalent bonds are formed when electrons are shared between atoms; the removal of an electron from a molecule normally weakens the bonding, causing an increase in bond length. This is illustrated by the comparison between the H_2 molecule and H_2^+ ion, given in table 7.5. But in the case of oxygen, the bond length actually decreases

Table 7.5 *Comparison of bonding in H_2 and H_2^+*

	H_2	H_2^+
Bond energy (kJ mol^{-1})	436	260
Bond length (nm)	0.074	0.11
Bonding electrons	2	1

when an electron is removed, forming O_2^+. This is simply explained by molecular orbital theory: the diagrams for O_2^+, O_2 and O_2^- are shown in figure 7.10. When oxygen loses an electron it is an anti-bonding electron which is lost, and so the resulting ion has more net bonding electrons

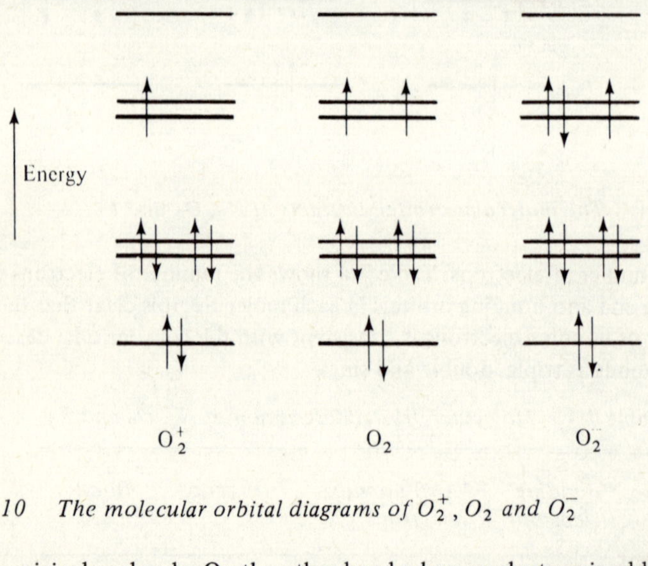

7.10 *The molecular orbital diagrams of O_2^+, O_2 and O_2^-*

than the original molecule. On the other hand when an electron is added to O_2, forming the superoxide ion, this goes into an anti-bonding orbital, thus weakening the covalent bond. Table 7.6 compares the covalent bonds in O_2^+, O_2 and O_2^-.

This brief section gives some idea of how molecular orbital theory can

Table 7.6 *Comparison of bonding in O_2^+, O_2 and O_2^-*

	O_2^+	O_2	O_2^-
Bond length (nm)	0.112	0.121	0.133
Net bonding electrons	6 – 1 = 5	6 – 2 = 4	6 – 3 = 3

describe the behaviour of electrons in molecules. In some cases it simply agrees with elementary valence theory, but in other cases, such as O_2, it makes new predictions of considerable accuracy. It is even capable of describing the bonding in radicals such as NO and NO_2 - but that is another story, and must be left for another time and another place.

Suggestions for further reading

Introduction to inorganic chemistry, G. I. Brown (Longmans)
Modern inorganic chemistry, C. Chambers and A. K. Holliday (Butterworth)
Investigation of molecular structure, B. Gilbert (Mills and Boon)
The Doomsday syndrome, J. Maddox (MacMillan)
Mechanisms in organic chemistry, R. O. C. Norman, M. J. Tomlinson and D. J. Waddington (Mills and Boon)
Air pollution, D. J. Spedding (Oxford Chemistry Series)
The chemist in action, Nuffield Advanced Series (Penguin)
New Scientist **81**, p. 474 (1979)

Index